Concepts of Inventions
Multiplexing Reality

©2013 John F. Roach

Foreword

In the importance to civilization, invention is underestimated. Much remains to be done with the work of invention in general.

Inventions vary in their social impact, some having little impact while others like the automobile transform society a great deal. This book deals with inventions or invention concepts that can mean a great deal in transforming society.

It is known that the world of invention is vast and wide. The products of technological invention not only include physical devices but also algorithms, processes, and biological structures.

Inventors sometimes respond to social needs by tackling already recognized problems. But sometimes they invent the problems themselves, discerning a problem that previously was not recognized as such. In this book both methods will be utilized, also in some cases small parts of many other inventions may be integrated to make one useful invention or an entirely different invention. In very rare occasions inventions will come to an inventor through a daydream or arbitrarily which may not be seen as useful at the time. Several of these will be presented in this book.

However even though this material is not necessarily in a patentable form, many concepts for inventions are revealed. It is therefore noted that much can be gleamed from its study.

"Sometimes, even the best ideas are filed on a shelf.

Good inventions are often not patented because of the process and expense. Prototypes are not built or, worse, stashed in a back room after the principle is proven. Theories are not always submitted for publication, or they are rejected because they are too radical, or too tangential to the main thrust of the publication to which they are submitted. Concepts of Inventions provide a voice to inventors whom otherwise may not be heard and a resource for those learning the process of getting their invention off the shelf."[1]

An example of concepts of inventions and inventions themselves can be demonstrated with the invention of Thomas Edison's light bulb. The concept of the light bulb was known long before Edison invented it. Edison looked far and wide for a variety of filaments for his invention.

It was known that if you could pass an electric current through some material and the material did not burn you could produce a light bulb. Edison eventually found a suitable filament and he got credit for the electric light, but the concept was necessary in order for him to successfully complete the invention. In other words Edison invented the filament, not the light bulb.

This book presents concepts of inventions. These are not necessarily the inventions themselves, but the concepts of the inventions. A total of 19 concepts of inventions are examined. Sometimes the concepts are very close to the inventions themselves and other times much research and development is necessary for the invention to be complete.

---

[1] http://www.inventionconcepts.com/

Contents

Foreword ...1

Storage Wheel and Battery Concept...4

Gyro Storage...10

Smart Grid Gyro Power...19

Two Improved Electric Motors...24

Hybrid Car Plus...33

Rotary Engine ...37

Retractable Studs...41

Solar House...43

Ethanol Still...49

Safe Nuclear Battery...53

Hybrid Water Heater...62

Switchless Pump...65

Jet Engine Booster...67

Fractal Stethoscope...72

Watch Voice Recorder...76

Liquid Gyroscope...78

Space Walk Propulsion...90

High Powered Pulsing Space Propulsion...92

Air Conditioned Suit...94

Section C1.Storage Wheel and Battery Concept

$$E = 1/2 \ Io \ V^2$$

Io is proportional to mass

velocity = force x time

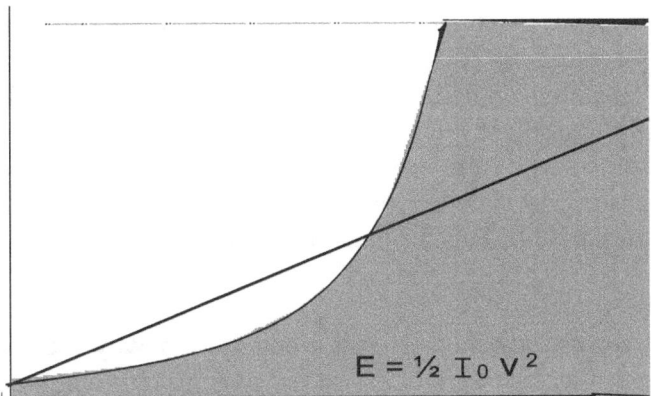

$$E = \frac{1}{2} I_0 V^2$$

The above diagram describes the kinetic energy under the V (velocity) curve with a constant **Io.**

The concept of the previous diagram illustrates that it is possible to store an infinite or more realistically a near infinite amount of energy in a rotating wheel in space. This will be the premise for all the kinetic storage devices in this book. In the diagram (E) represents energy, Io represents a type of mass or circular structure and (V) represents velocity in the form of angular velocity and is the velocity at the rim of the wheel.

$$\mathbf{v}_i = \omega \times \Delta\mathbf{r}_i + \mathbf{V}_R$$

Where ω is the angular velocity of the system, and $\mathbf{V}_R$ is the velocity of **R** and Where **V** will be simply referred to as the rotational velocity of the wheel. For simplicity V is used.

Take for instance the rotating wheel. For purposes of illustration, a rotating wheel exists in space where there is no friction and nothing to retard its motion. The wheel is fed a steady force or push. Never mind how it receives this steady push, we are using our imagination and can do anything to the wheel we want.

As long as the wheel receives an incremental push of force in space the velocity of the rotating wheel will increase. When the velocity of the rim of the wheel approaches the speed of light it will go no faster. When this happens the energy stored in the wheel will be virtually infinite.

To use this energy in a practical manner we must bring it down to earth, so to speak. We must literally harness it to a physical structure. The structure will consist of 2 resistive points, Ball Barings at the top and bottom. This is the maximum configuration. It is possible to configure gears and possibly governors of some sort, but in order to maximize the structure we will refrain from doing so. The Ball Barings add a resistive or retardant force that should be minimized. This is done by using expensive least resistive components.

You might ask should we use a battery to store energy and why are we using a rotating wheel to store energy since a battery has no moving parts and with enough batteries we could theoretically store as much energy as the kinetic device. The reason is that a battery cannot absorb energy as quickly as a rotating wheel. The rotating wheel can discharge instantly which if configured properly will yield electrical or mechanical energy. Of course if the energy is not recovered when it is available, it is lost, however with low wheel friction the energy is available for a relatively long time.
In the real world there are limitations. We cannot obtain the non-friction environment of space and the wheel must be of reasonable size. The velocity of the wheel although it will be maximized will never approach the speed of light.

Therefore we cannot have a perfect kinetic storage device, but it will be shown that a kinetic energy device is far superior to other storage systems for short durations and high power. Some of the limitations of the system can be supplemented with batteries to provide more efficient storage however the particular kinetic system described above is in need of great improvement at this time having little or no use except in space. We will use a battery to give the device a longer discharge time.

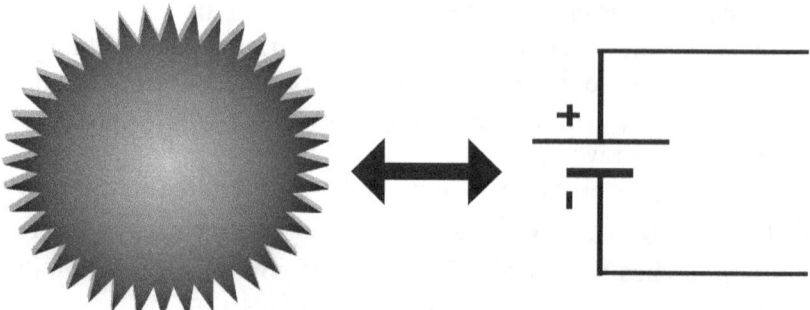

The problem with batteries alone is their inability to accept a charge quickly enough. This is because there must be a change in the molecular structure of the battery. This physical change is relatively slow.

Within the realms of practicality for a rotating system there are really only a few things that prevent us from having a better storage system for use in the gravity of earth and they are: friction, reasonable size and strength, endurance of ball bearings and endurance of the electrical system. The cost of course is a consideration because more durable parts are more expensive, but longer life and less maintenance would be achieved.

Many of the systems of the past have not had the technology to include electronic gearing that is providing pulses of electrical energy to the wheel they've had to use mechanical gears. These mechanical gears waste a considerable amount of energy. Approximately 30% of the energy will be wasted in the transmission from the wheel to the mechanical gears.

In our system we will introduce the energy with electromagnetic fields and technology such as advanced micro-controllers. We will receive the energy likewise with technology and micro-controllers.

Another way to improve our system is to encapsulate the rotating wheel inside a vacuum. There is a tremendous amount of friction in the form of drag when the wheel rotates in a gas this comes from an effect which has been studied in aerodynamics this is a particular type of resistance known as the Coanda effect. A tremendous amount of gain results from putting the rotating wheel inside a vacuum. In fluid-dynamics a gas such as air is considered a fluid.

"The Coandă effect is a result of entrainment of ambient fluid around the fluid jet. When a nearby wall does not allow the surrounding fluid to be pulled inwards towards the jet (i.e. to be entrained), the jet moves towards the wall instead. The fluid of the jet and the surrounding fluid should be essentially the same substance (a gas jet into a body of gas or a liquid jet into a body of liquid). In one application, a jet of air is blown over the upper surface."[2]

"The Coanda effect or wall-attachment effect, is the tendency of a moving fluid, **either liquid or gas**, to attach itself to a surface and flow along it. As a fluid moves across a surface a certain amount of friction (called "skin friction") occurs between the fluid and the surface, which tends to slow the moving fluid. This resistance to the flow of the fluid pulls the fluid towards the surface, causing it stick to the surface. Thus, a fluid emerging from a nozzle tends to follow a nearby **curved surface- even to the point of bending around corners**-if the curvature of the surface or the angle the surface makes with the stream is not too sharp. Discovered in 1930 by HenriCoanda,Romanian aircraft engineer, the phenomenon has many practical applications."[3]

[2] http://en.wikipedia.org/wiki/Coand%C4%83_effect

[3] http://www.boomerangs.com/airfoils.html

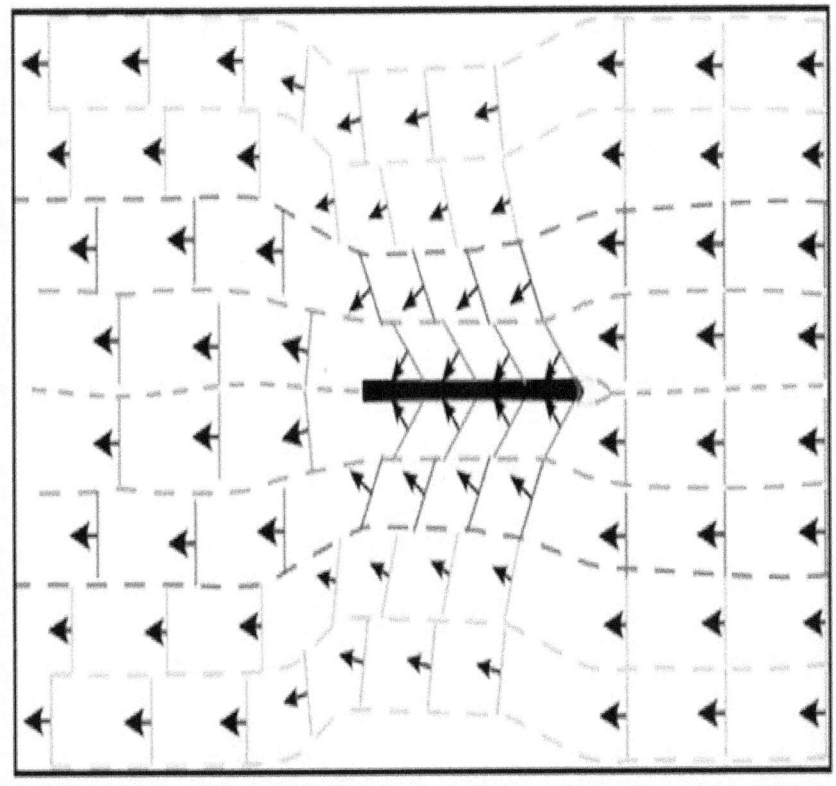

Note 1. The I   is a fluid stream front

Note 2. There are 2 types of friction:

a) Friction between the lowest air stream and the foil. The friction bends the layer front into foil surface. This results in "dynamic sticking to the surface".

b) Friction between different layers of streams (dynamic viscosity). The lower layer of the stream drags and bends down the upper neighboring layer.

Note 3. The stream layers have velocities gradient along the normal to foil surfaces.

Note 4. The figure is simplified - the foil is considered to be very thin. Air flow is laminar. The stream bending (and little compression for air only) is exaggerated in order to show it.

Note 5. The Coanda effect has a dynamic origin - it appears at nonzero relative to fluid and foil speed_

Section C2. **Gyro Storage Small/Medium Scale**

This section will discuss the operation of a small/medium rotational storage device. Figure C-2.1 shows the bare bones of a small rotational storage system. The wheel may be rotated either clockwise or counterclockwise, but at all times must be contained within a vacuum.

The spherical Neodymium magnets are extremely powerful and will be used with all of the medium and small rotational systems.

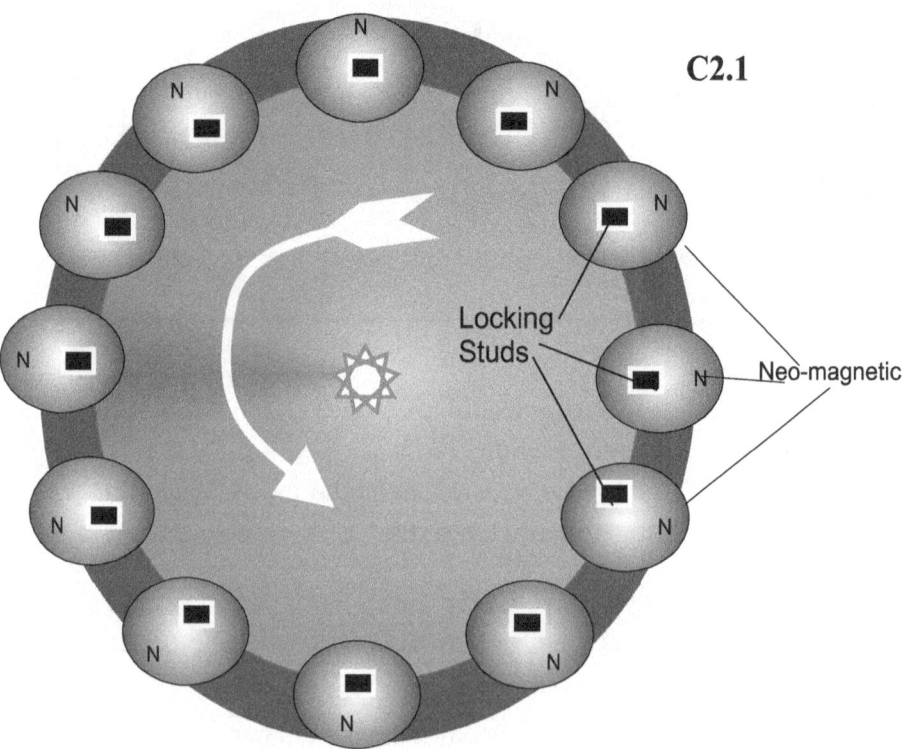

Because of the relatively small radius of the wheel all the Neo magnets should be spherical in small to medium devices.

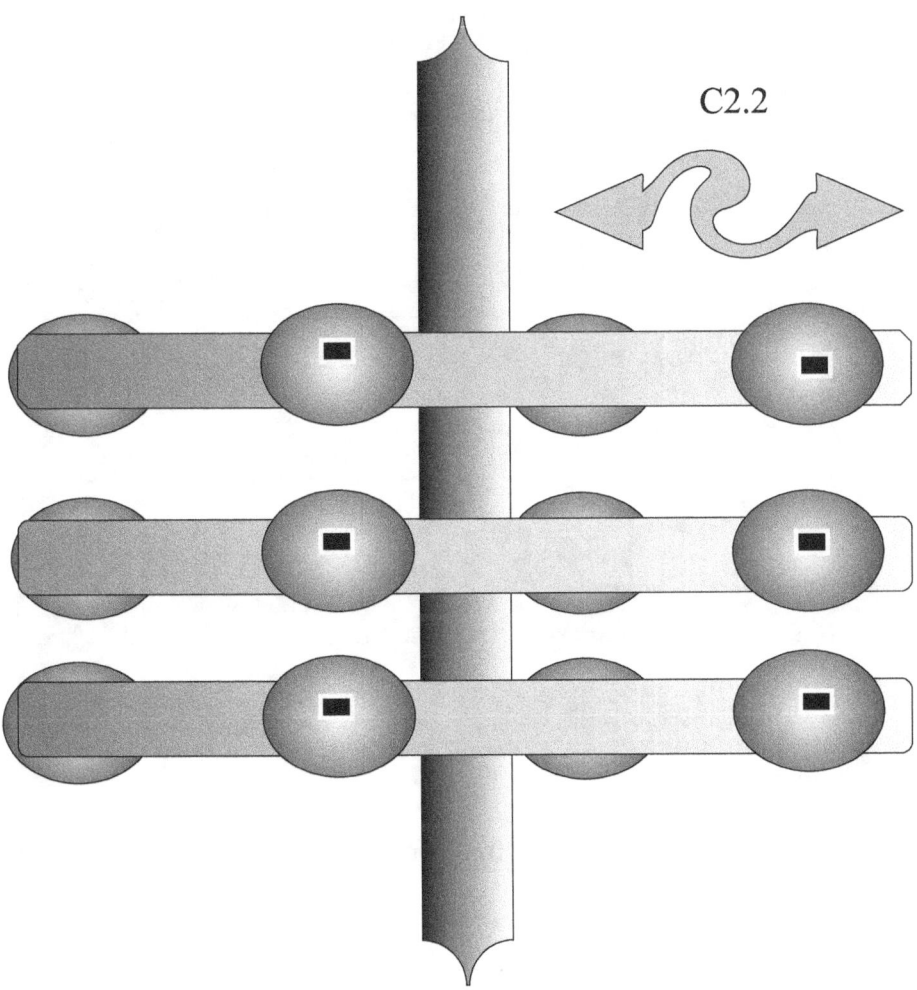

Figure C-2.2 is a demonstration of how to increase the angular mass without increasing the diameter of the wheel. It is important to have a rather large angular mass even though the velocity produces the absolute energy of the system. The angular mass is the time it takes to charge and discharge the energy, thus allowing more time to charge the battery. Another way to increase the angular mass is to increase the diameter of the wheel. A small increase in the diameter will result in a large increase in the angular mass or Io. To compute kinetic energy I0 for all practical purposes may be considered to be the mass of a non-rotating body.

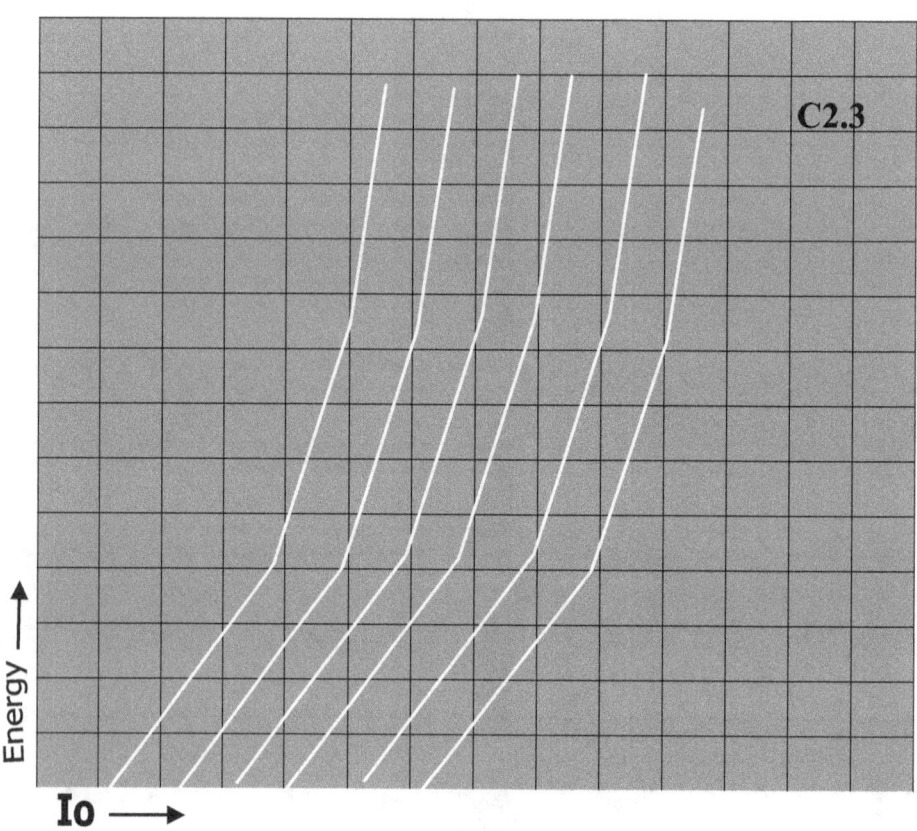

The above figure illustrates resulting energies when using different **Io** values.

"When a body is rotating around an axis, a torque must be applied to change its angular momentum. The amount of torque needed for any given change in angular momentum is proportional to the size of that change. The constant of proportionality is a property of the body that combines its mass and its shape, known as the moment of inertia. In classical mechanics, **moment of inertia** may also be called **mass moment of inertia**, **rotational inertia**, **polar moment of inertia**, or the **angular mass** (SI units $kg \cdot m^2$, US units $lb_m \, ft^2$)."[4]

---

[4] http://en.wikipedia.org/wiki/Moment_of_inertia

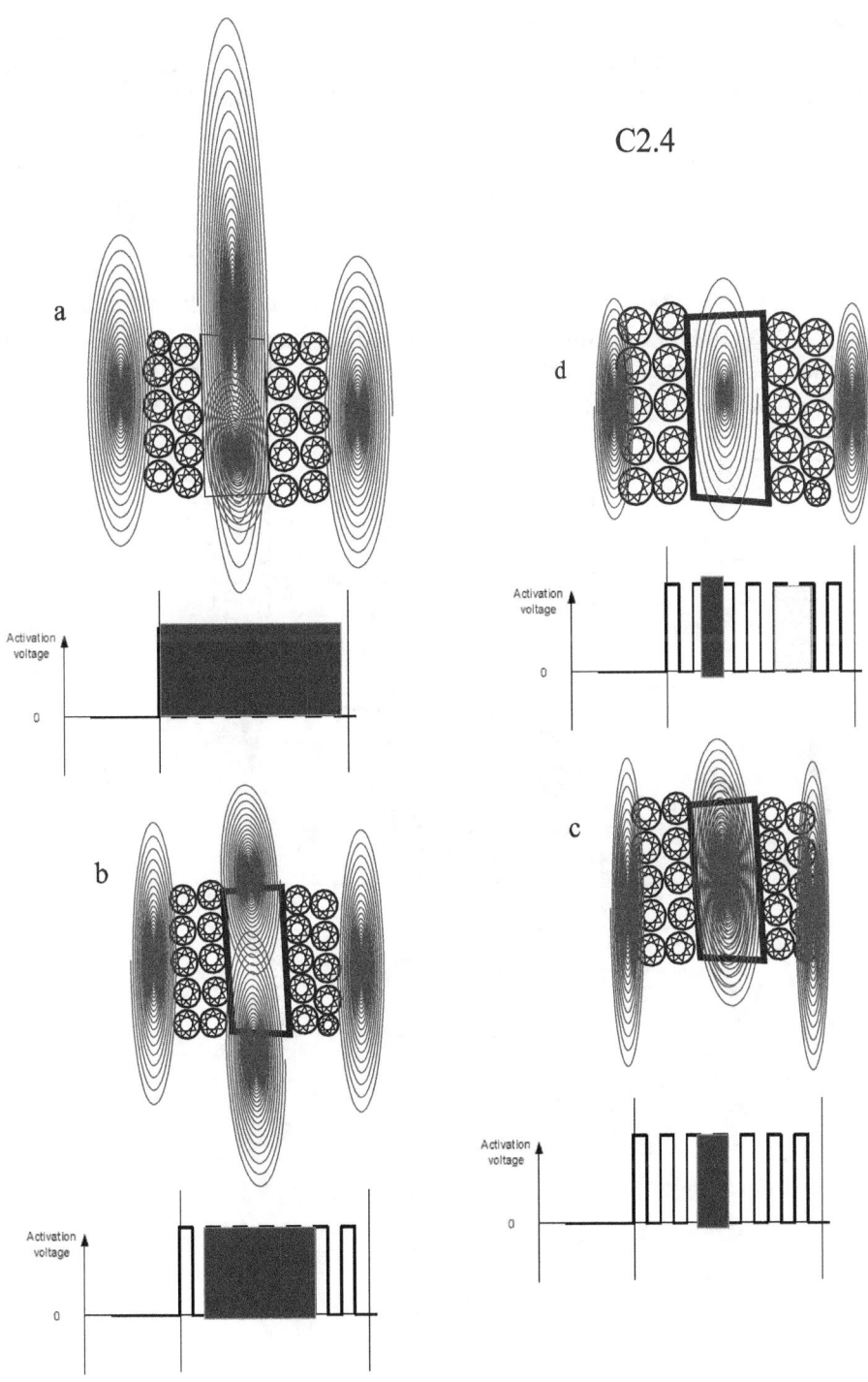

C2.4

The excitation of the inductors can be seen in figure C-2.4. In part (a) of figure C-2.4 the maximum amplitude signal is being applied to the coil. You can see from the figure that pulse width modulation is being used. In part (b) a lesser amount of signal is being applied and in part (c) and (d) lesser amounts of signal are being applied. The frequency of the pulse width modulation should be above the audible frequency that is above 20,000 Hz.

Several of these type systems are being used to propel the wheel. These systems must balance in other words you could have 2, 4, 6 or 8 or some combination of these. As long as they were even. This is necessary for the wheel to balance properly.

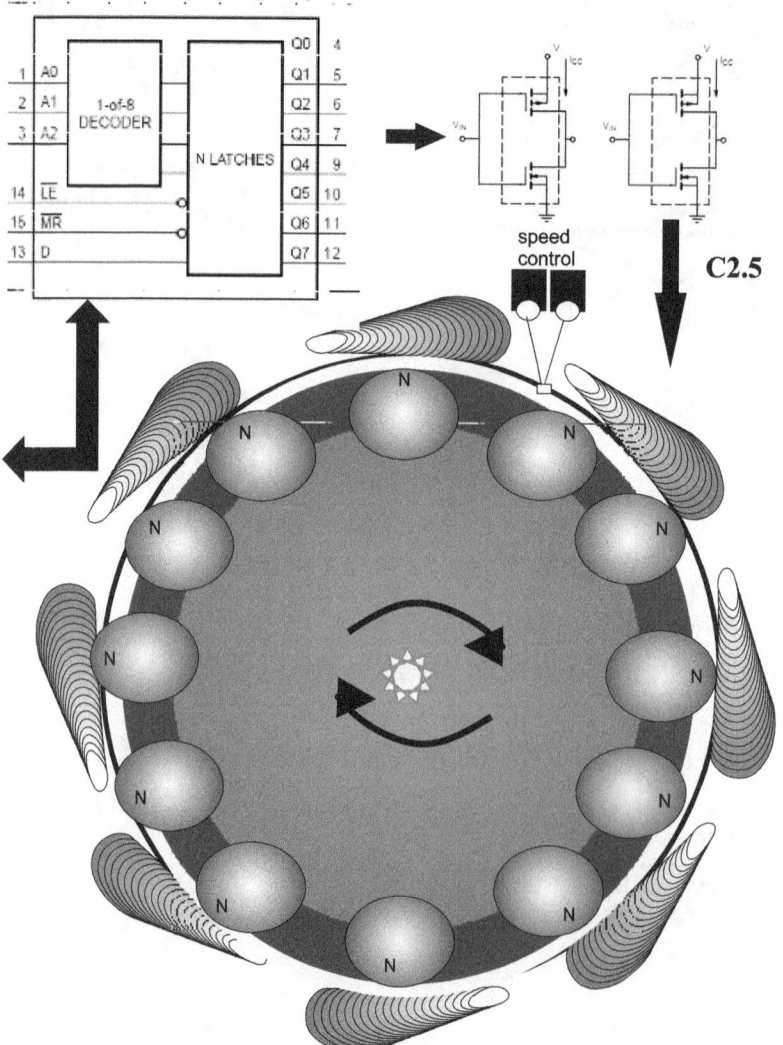

Figure C2.5 shows the key parts of the rotational storage system minus the coils and microprocessor.

Figure C2.6 shows the system with the coils and speed detector that goes to the microcontroller. The speed detector is used to determine pulse width modulation.

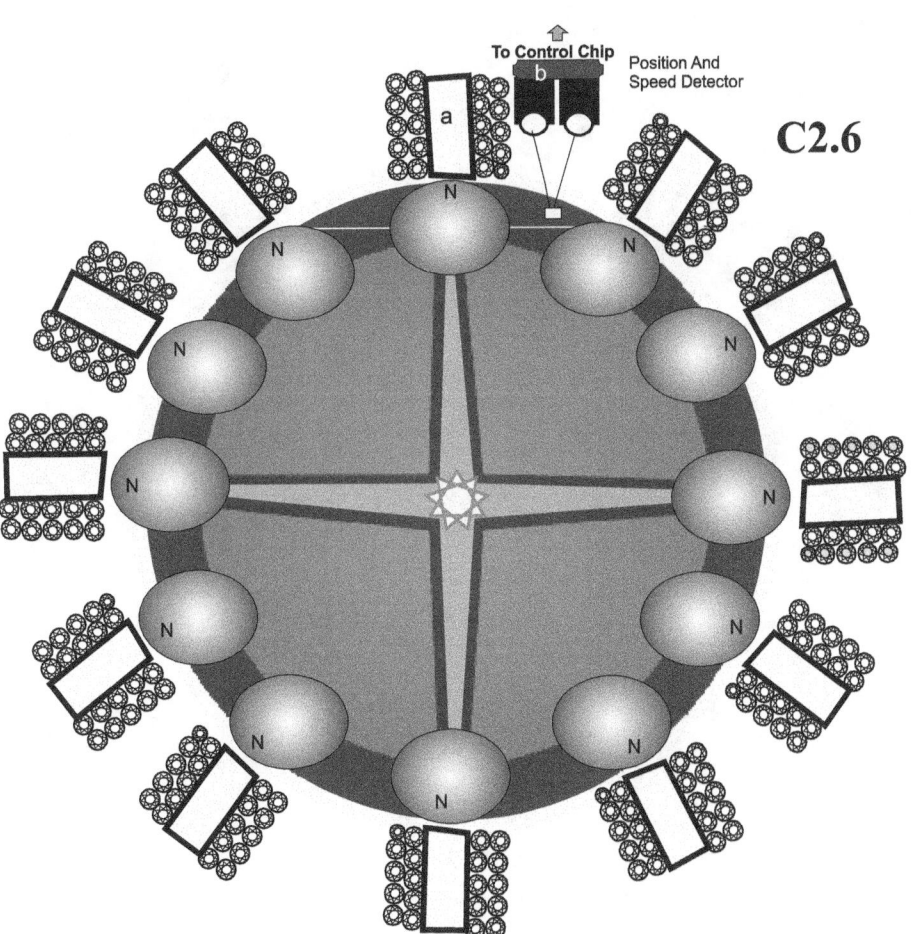

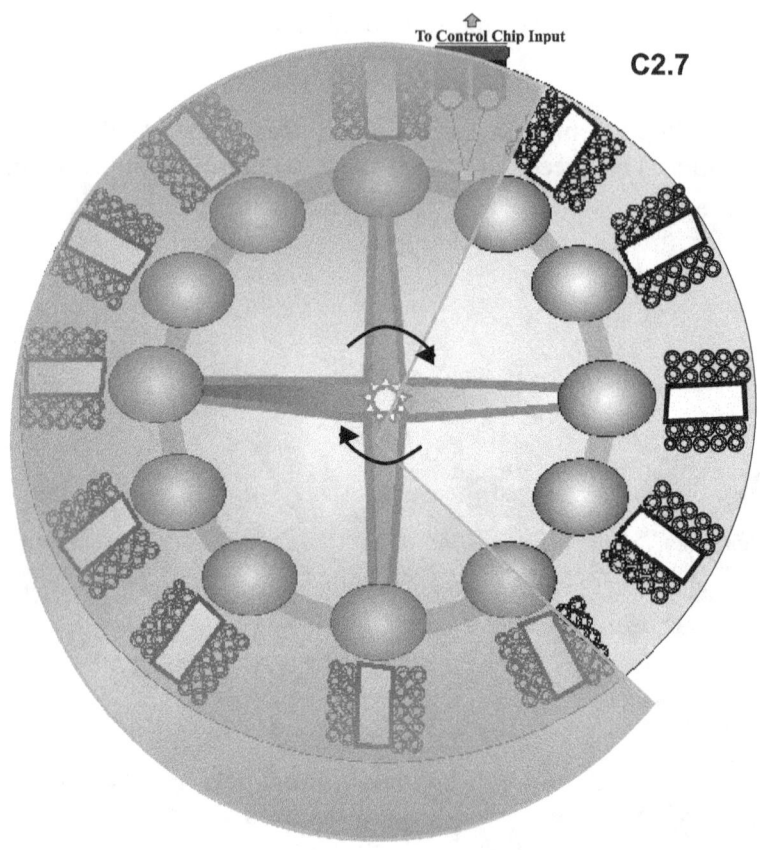

C2.7

All input currents connect from the Output of the CMOS separately.

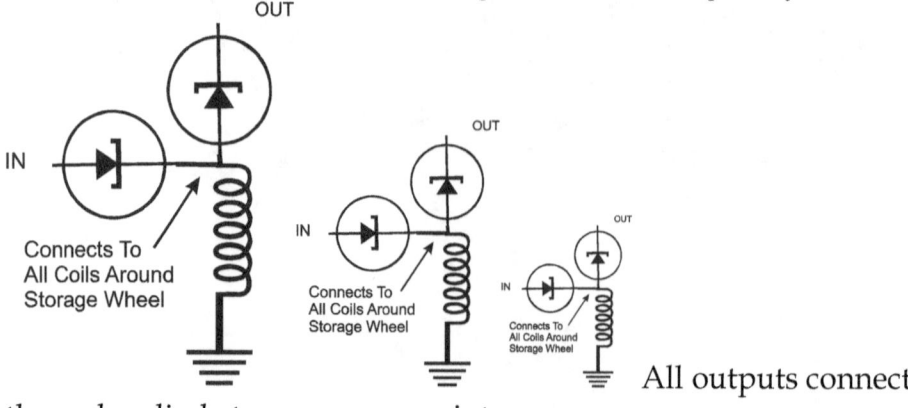

through a diode to a common point. All outputs connect

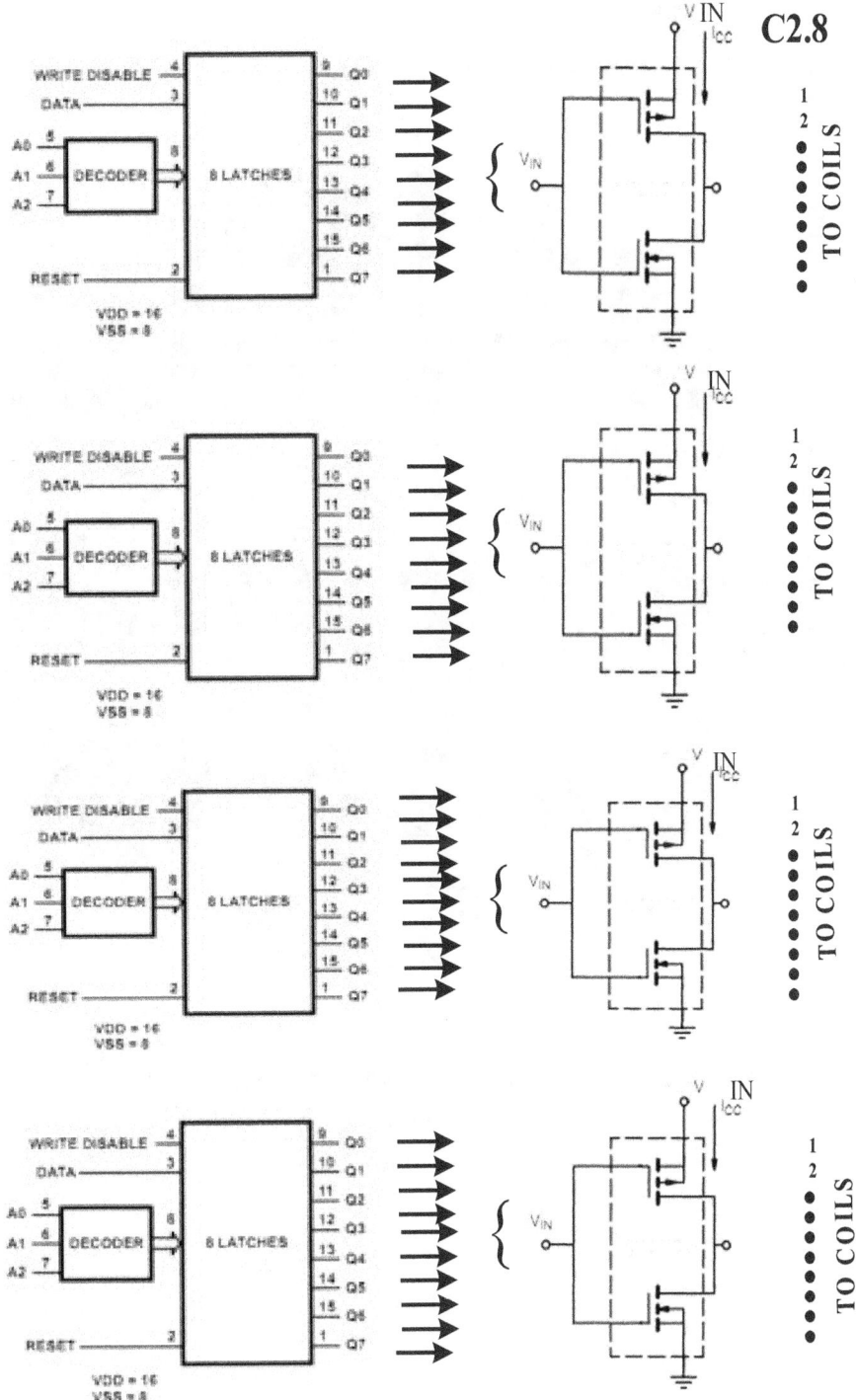

C2.8

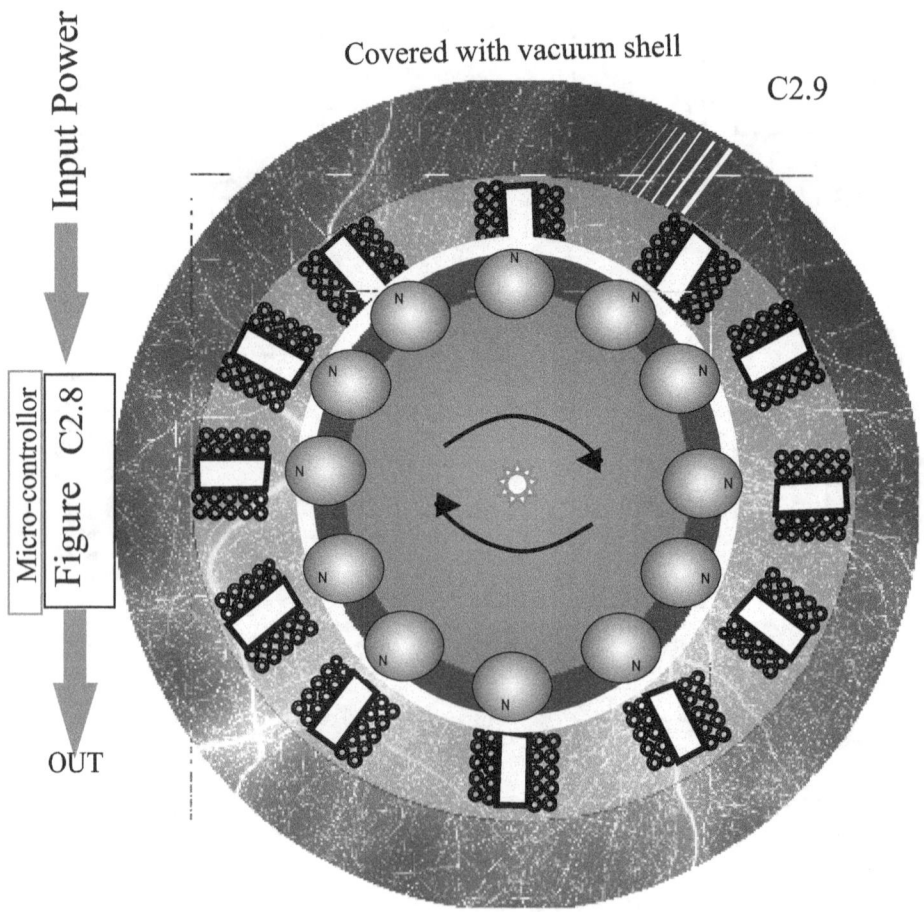

Figure C2.9 shows the spinning wheel in operation. That is it shows the magnetic and electric fields. These EMF fields can be used to power a small balancing device if necessary.

C3.Smart Grid Gyro Power

"Heat causes increase loss   better to be closer to destination
Power outages caused by warmer days in the United States would necessitate current electricity infrastructure to be equipped with smart grid technologies, a power grids executive said.

"As record-breaking temperatures in the Northeastern United States have consumers turning up their air conditioners, our overburdened electric grid is taking center stage," said Mr. Gilligan.

"The heat wave is causing spikes in power demand, increasing the probability of service disruptions," he added.

Power service disruptions have been reported recently in Massachusetts, California and Michigan, among others. The Boston Globe recently reported that approximately 4,000 of utility Nester's electric consumers suffered from power outages across the company's service region, which includes the city of Boston.

Over 2,500 customers of Eastern United States electric company National Grid also experienced service disruptions. Another 2,000 customers of the seaside community Laguna Beach suffered from a power outage recently, the Orange County Register in California said.

The Chicago Tribune also reported that pockets of downtown Detroit were left without electricity as power outages struck the area on Saturday.

Caroline Allen, a spokeswoman for OnStar, told the Boston Globe that transmission lines were subjected to tremendous stress for several consecutive days. She said that such problems with the power grid were expected following several days of extensive electricity consumption..

Smart grid technologies can equip power grids with demand response support. This feature allows generators to manage spikes in power demand more intelligently.

"Investing in smart grid technologies can enable utilities to better manage fluctuating consumer demand and keep the [air conditioning] on without adding more generation and transmission capacity," Mr. Gilligan said."[1]

See section C2 for detailed construction of gyro wheel storage circuits.

The wheel is rotated and stores kinetic energy, which is turned into electrical power.

---

[1]   http://www.xing.com/net/erneuerbareenergien/new-technologies-neue-technologien-477/smart-grid-the-answer-to-power-demand-spikes-says-g-e-executive-31243719

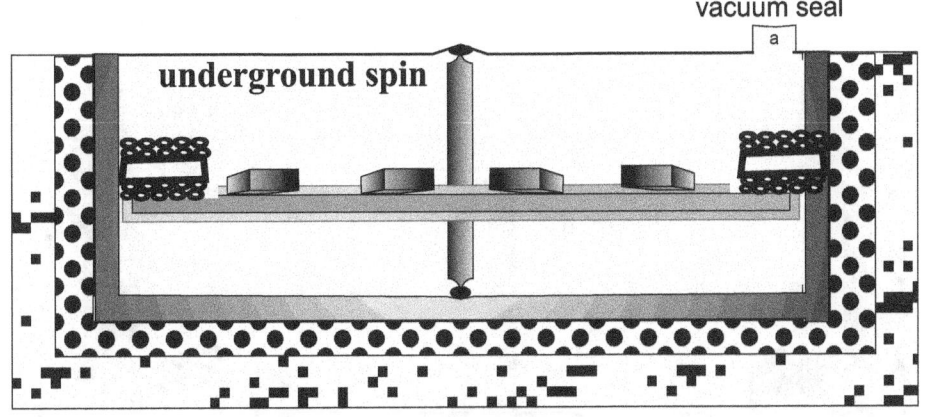

vacuum seal
a
underground spin

C3.1

Coils
Magnets
a

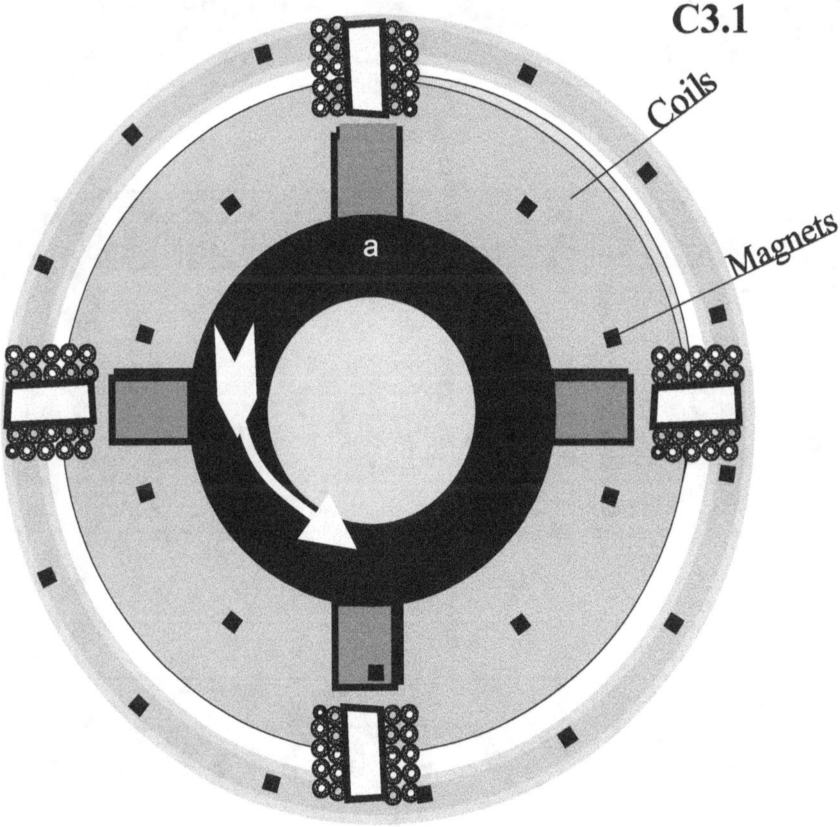

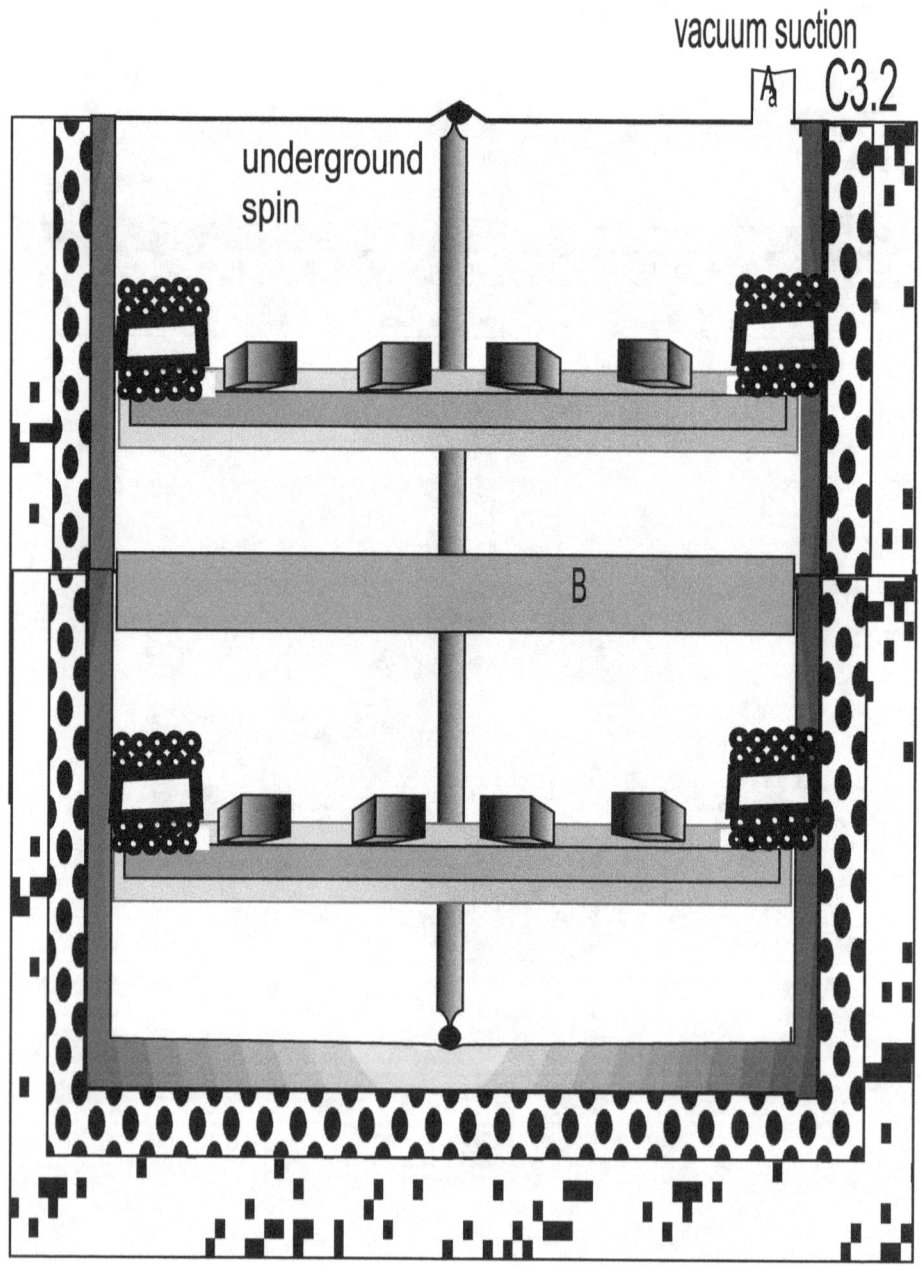

Figure C-3.1 will add a buffer and regulation for rural locations and figure C-3.2 will add buffer and regulation for urban locations.

Either one of these two systems can be used to prevent a relatively short power blackout if used in conjunction with a battery bank.

Load balancing

"The transmission system provides for base load and peak load capability, with safety and fault tolerance margins. The peak load times vary by region largely due to the industry mix. In very hot and very cold climates home air conditioning and heating loads have an effect on the overall load. They are typically highest in the late afternoon in the hottest part of the year and in mid-mornings and mid-evenings in the coldest part of the year. This makes the power requirements vary by the season and the time of day. Distribution system designs always take the base load and the peak load into consideration.

The transmission system usually does not have a large buffering capability to match the loads with the generation. Thus generation has to be kept matched to the load, to prevent overloading failures of the generation equipment. In centralized power generation, only local control of generation is necessary, and it involves synchronization of the generation units, to prevent large transients and overload conditions."[2]

---

[2] http://www.science.gov/topicpages/u/ubiquitous+power+grid.html

Section C4. Two Improved Electric Motors

An electrical motor concept for two electrical motors is presented. The first electrical motor will take advantage of two relatively new inventions. One of the inventions is the extremely compact and inexpensive microcontroller which allows for any contingency especially those given in the information reference section. Another is the tremendous power of Neodymium magnets which allow for much greater power from their small size. Neodymium magnets are the most powerful permanent magnets on earth, and are just now being used for certain applications.

Another is the use of pulse width modulation. By using pulse width modulation it is possible to obtain analog emulated signals. In figure C4.1 it can be seen how each of these analog simulated pulses can move around the rotor of the electric motor.

Figure C4.2 shows the pulse width modulation that would be applied to all for coils. This allows for any compensation and torque reduction. In case of the motor stall the micro controller would be able to present different algorithms in an attempt to solve the problem.

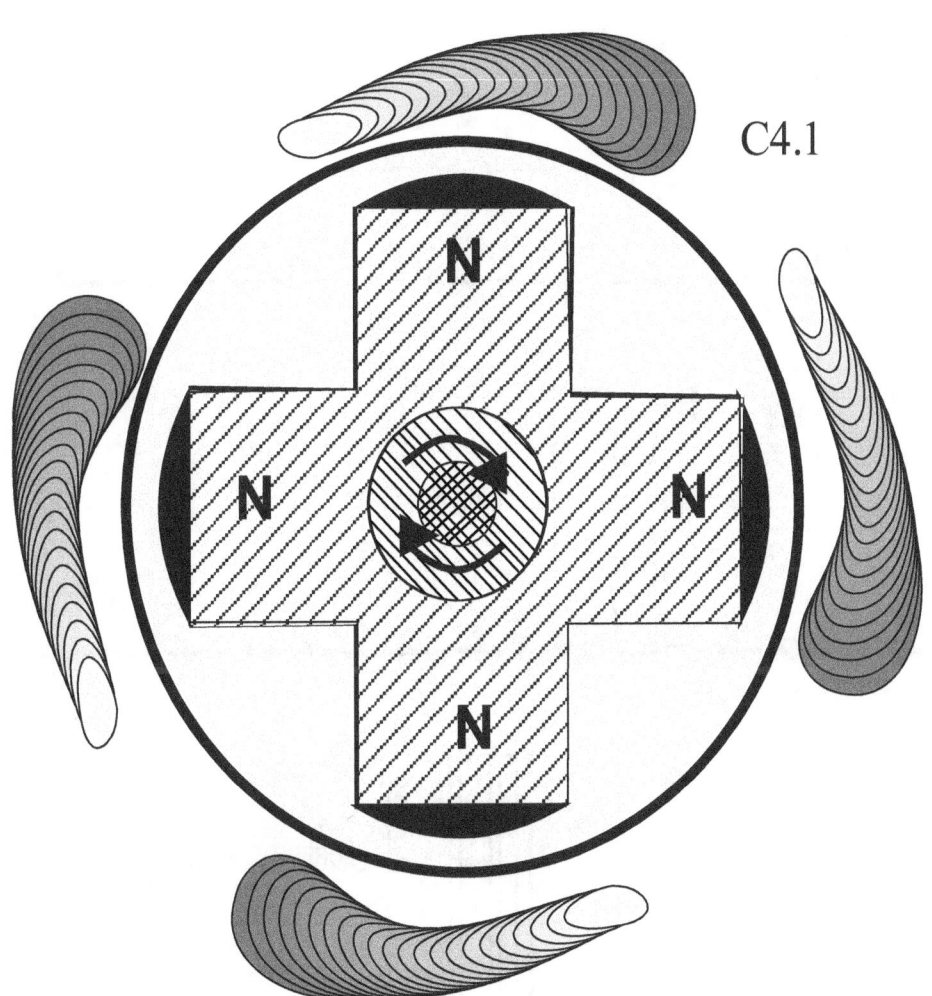

C4.1

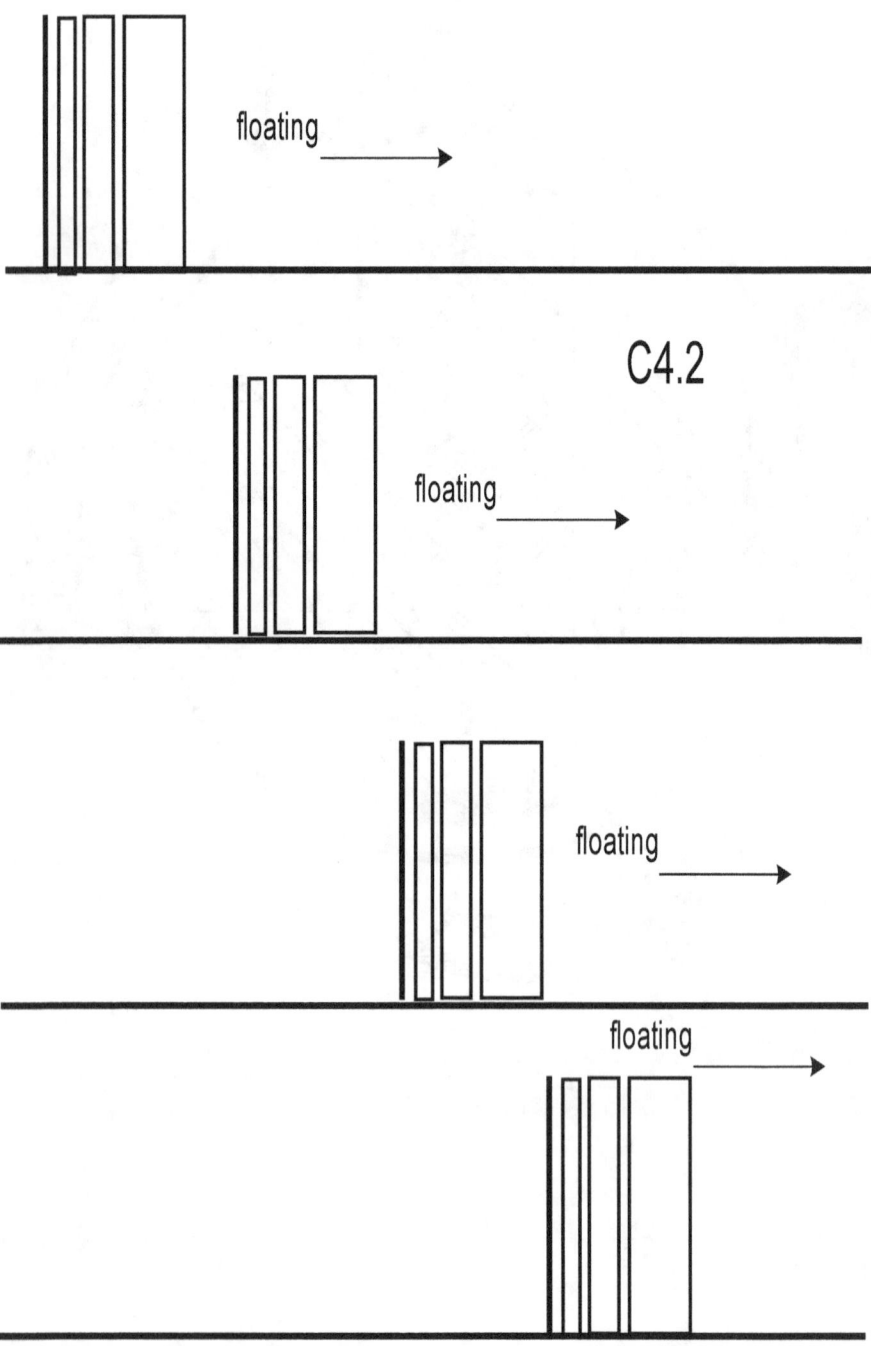

C4.2

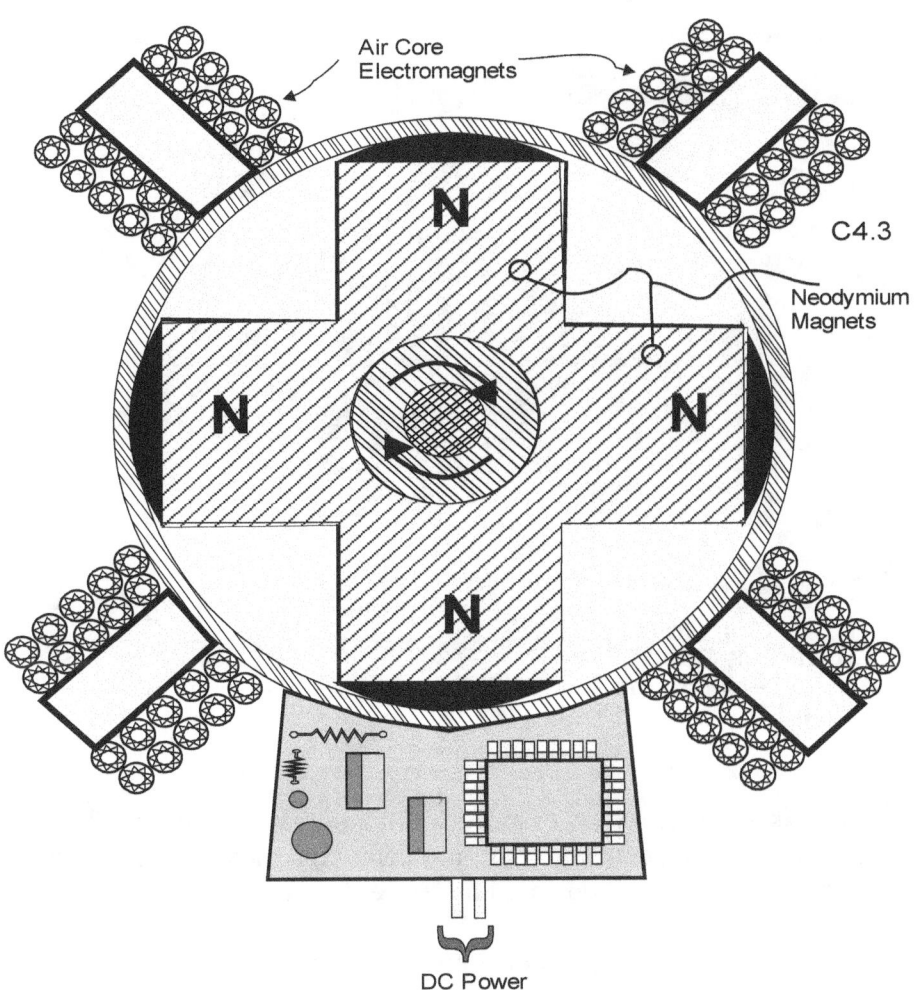

Side cut of pulse modulation motor with micro controller and Neodymium magnets.

Any problems that this motor concept has, should be eliminated and are at least addressed under reference information. In a conventional motor torque is small at some angles. This will not happen with these motors.

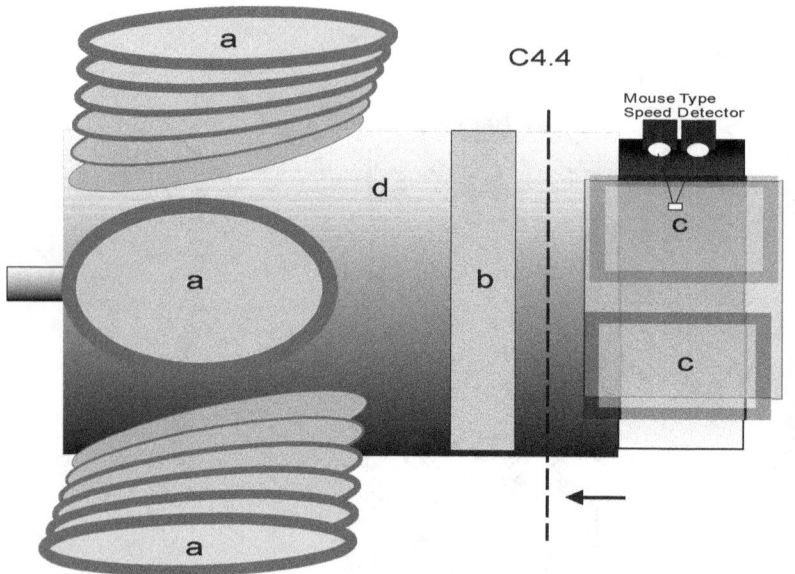

C4.4

Mouse Type
Speed Detector

a Air Coil Coils
b Circuit Board
c Current Drivers For Motor
d Nonmagnetic Shell

This motor is a Pictorial view of pulse width modulation with speed control and temperature compensation. Any problems this motor concept should be eliminated, are at least addressed under reference information.

Another improved electrical motor concept is presented. This electrical motor will take advantage of a shaft connection to another motor section and also a microchip that will measure the rotational speed of the shaft. The shafts velocity can be used to increase the low level of torque that is present in a brush motor it will also provide tachometer information that may be used for some other purpose.

A low level LED light will be used to gain information from the bar-coded shaft. This information will be used by the microchip to send the appropriate voltage to the brush motor.
Any number of chambers can be used to extend the information to other sections of the brush motor however to prevent diminishing

returns N should be no greater than one. The figures for these two motors are shown in C4.5 and C4.6 respectively.

One other advantage of this type of motor could be seen in RC racing cars. High-speed would be exchanged for lower lifetime.

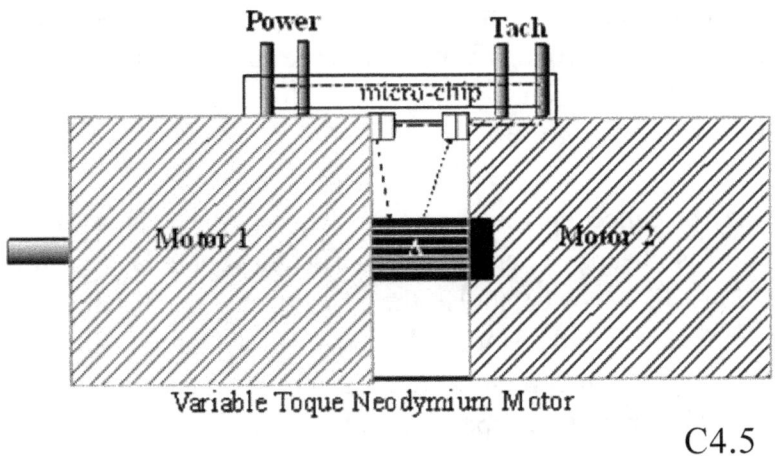

Variable Toque Neodymium Motor

C4.5

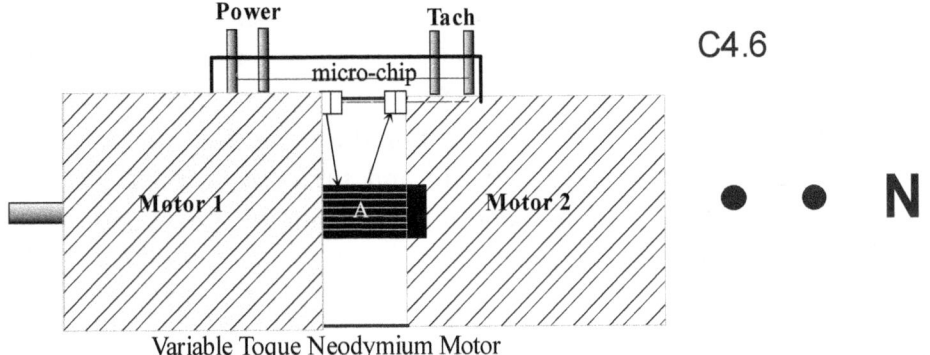

Variable Toque Neodymium Motor

C4.6

Referenced information section:

"Single phase induction motors have problems for applications combining high power and flexible load conditions. The problem lies in producing the rotating field. Shaded poles are used instead, but the torque is small at some angles. If one cannot produce a smoothly rotating field, and if the load 'slips' well behind the field, then the torque falls or even reverses.

Power tools and some appliances use brushed AC motors. Brushes introduce losses (plus arcing and ozone production). The stator polarities are reversed 100 times a second. Even if the core material is chosen to minimize hysteresis losses ('iron losses'), this contributes to inefficiency, and to the possibility of overheating.

These motors may be called 'universal' motors because they can operate on DC. This solution is cheap, but crude and inefficient. For relatively low power applications like power tools, the inefficiency is usually not economically important.If only single phase AC is available, one may rectify the AC and use a DC motor. High current rectifiers used to be expensive, but are becoming less expensive and more widely used. "[3]

"For all electrical motors the problem of the motor stalling (being locked) must be considered — if not. the motor can be permanently damaged or even take fire. Therefore any implementation of motor control must be able to respond to a stall situation. In some cases the stall is temporary. and it can thus be desired to restart the motor. A common approach is to stop the commutation of the motor when stalling is detected, wait for a while and then try to restart it. A microcontroller based motor control can handle this easily by monitoring the rotation speed.

A potential problem for an electrical motor and a driver stage is that it overheats. This occurs if the motor draws high currents. e.g. due to driving a large load or accelerates fast. If a temperature sensor is

---

[3] http://www.animations.physics.unsw.edu.au/jw/electricmotors.html

embedded in the motor or the driving stage, the ADC of a Most common is it to monitor the current flow"[4]

"Was originally written to help high school students and teachers in New South Wales, Australia, where a new syllabus concentrating on the history and applications of physics, at the expense of physics itself, has been introduced. The new syllabus, in one of the dot points, has this puzzling requirement: "explain that AC motors usually produce low power and relate this to their use in power tools.

AC motors are used for high power applications whenever it is possible. Three phase AC induction motors are widely used for high power applications, including heavy industry. However, such motors are unsuitable if multiphase is unavailable, or difficult to deliver. Electric trains are an example: it is easier to build power lines and pantographs if one only needs one active conductor, so this usually carries DC, and many train motors are DC. However, because of the disadvantages of DC for high power, more modern trains convert the DC into AC and then run three phase motors.Single phase induction motors have problems for applications combining high power and flexible load conditions. The problem lies in producing the rotating field. A capacitor could be used to put the current in one set of coils ahead, but high value, high voltage capacitors are expensive. Shaded poles are used instead, but the torque is small at some angles. If one cannot produce a smoothly rotating field, and if the load 'slips' well behind the field, then the torque falls or even reverses.

Power tools and some appliances use brushed AC motors. Brushes

---

[4]http://www.datasheetarchive.com/AVR440/Datasheet-025/DSA00435567.html

introduce losses (plus arcing and ozone production). The stator polarities are reversed 100 times a second. Even if the core material is chosen to minimise hysteresis losses ('iron losses'), this contributes to inefficiency, and to the possibility of overheating. These motors may be called 'universal' because they can operate on DC. This solution is cheap, but crude and inefficient. For relatively low power applications like power tools, the inefficiency is usually not economically important. If only single phase AC is available, one may rectify the AC and use a DC motor. High current rectifiers used to be expensive, but are becoming less expensive and more widely used. If you are confident you understand the principles, it's time to go to How real electric motors work by John Storey. Or else continue here to find out about loudspeakers and transformers."[5]

[5] http://electricalmachinery.blogspot.com/2010/11/some-notes-about-ac-and-dc-motors-for.html

Section C5.Hybrid Car Plus

This is the Hybrid Car Plus concept. For more information on the details of the storage wheel see section C2. This invented concept will allow an automobile to get more gas mileage and at the same time allow the car can be much bigger. The storage system in section C2 is used in conjunction with a battery and the car wheels are run completely on electricity.

The car or SUV will use electric shocks as well as reverse breaking also power on a decline surface. Figure C-5.1 shows an exposure of the main power wheel or storage device.

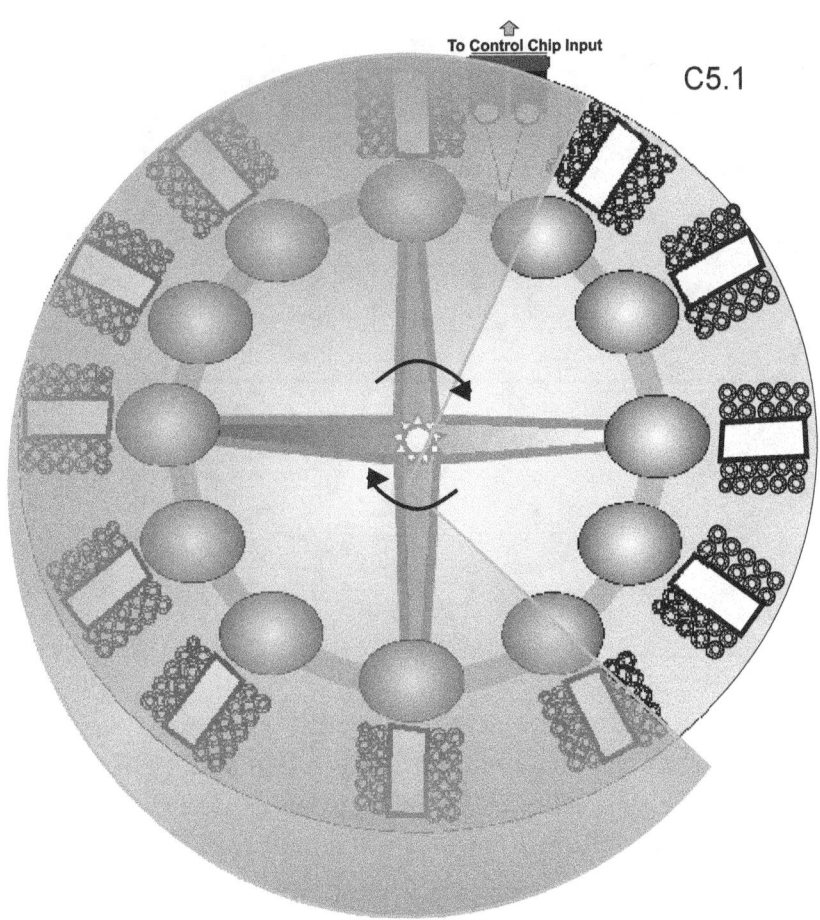

To Control Chip Input

C5.1

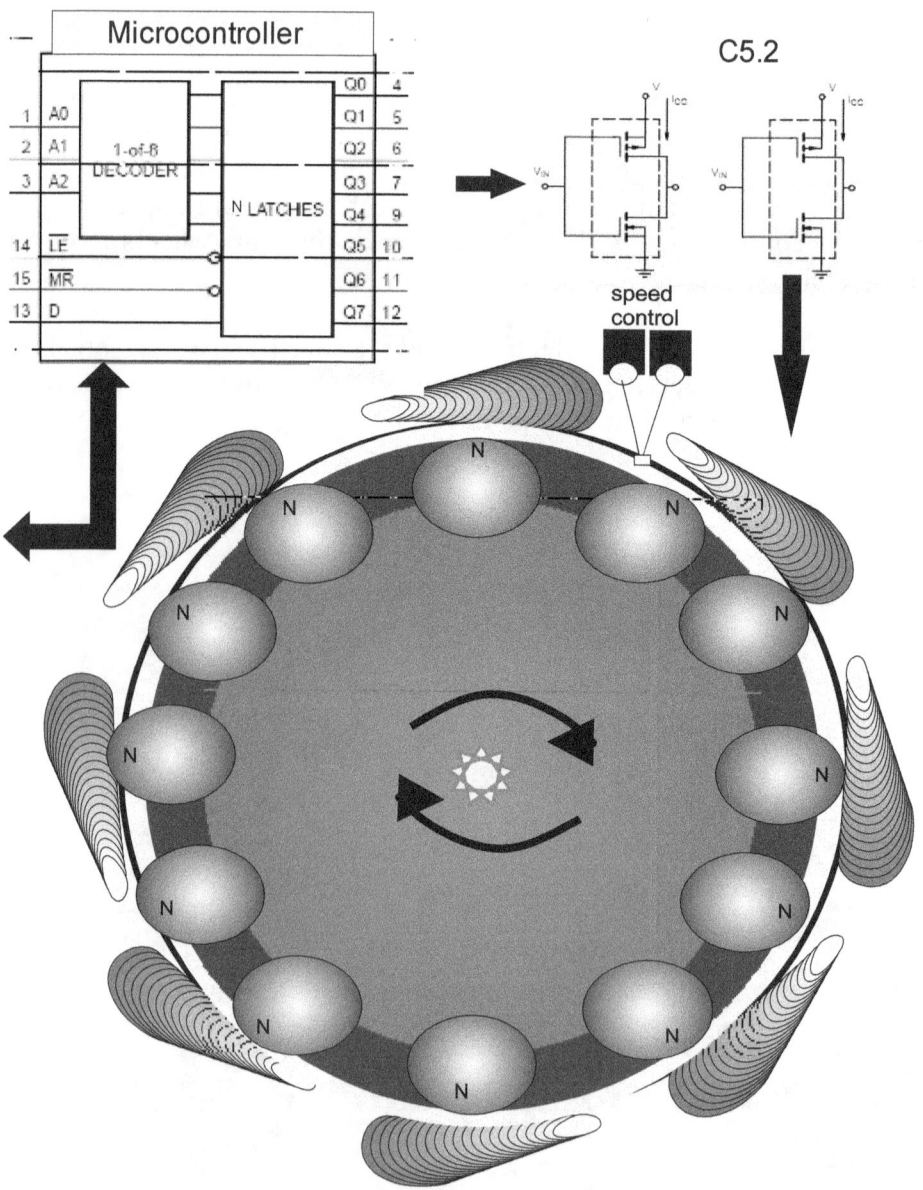

Figure C5.2 shows the basic electronics of the storage wheel. The wheel can spin up much faster than a battery can charge. That is it can hold a much larger amount of energy for a short amount of time and give it back to the battery when it can accept it.

All of the energy is sent to the storage wheel until it is "charged". Then it can be used to charge the battery and run the electric motor in the car. Only a small amount of fuel is needed to bring the storage wheel to the optimal value.

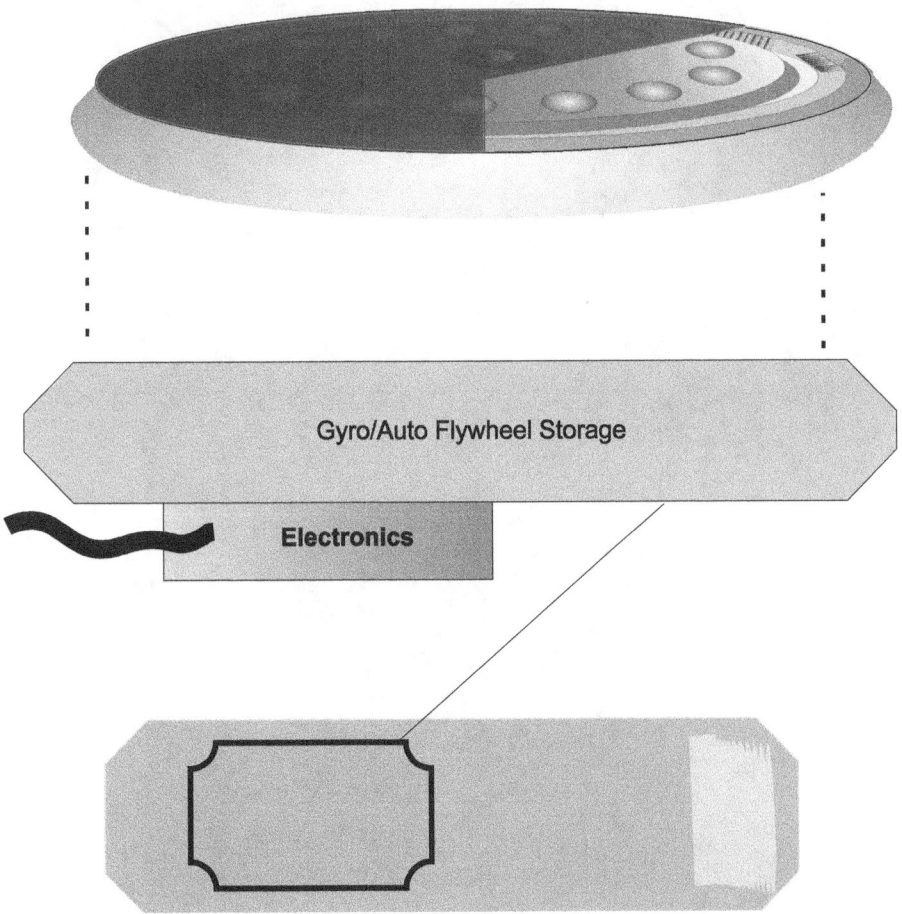

The Gyro/Auto vacuum Flywheel Storage requires a special suspension and as shown in the Figure can be located under the frame.

This special suspension can be monitored by a microprocessor or microcontroller and adjusted for any compensation in motion.

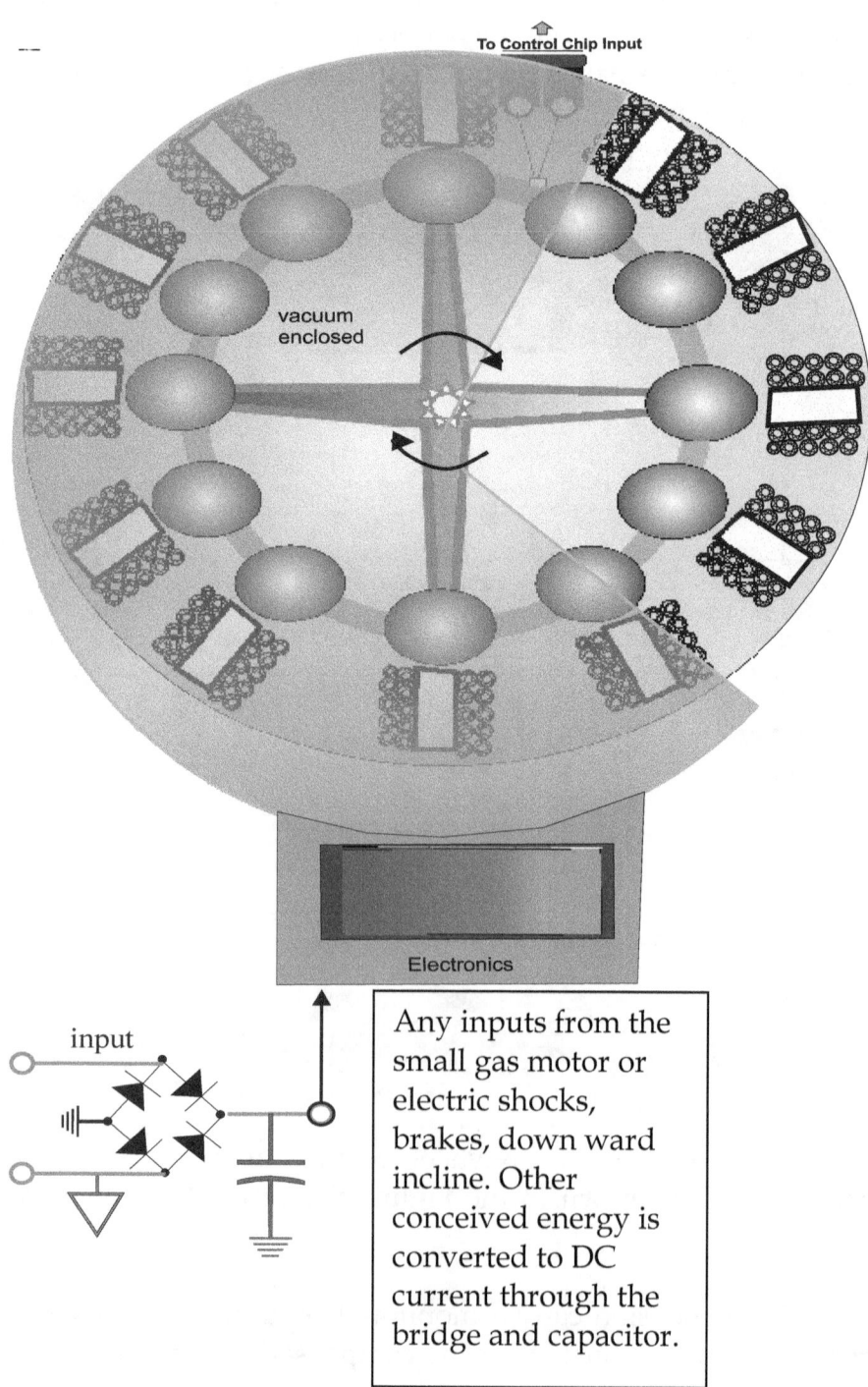

To Control Chip Input

vacuum
enclosed

Electronics

input

Any inputs from the
small gas motor or
electric shocks,
brakes, down ward
incline. Other
conceived energy is
converted to DC
current through the
bridge and capacitor.

Section C6. **Rotary Engine**

This concept of invention reveals a new and radically different type of Rotary engine. Because the engine is so radically different it would require a large amount of capital and development, but has the advantages an internal combustion engine would not possess.

If one examines the modern-day cylinder internal combustion engine it is apparent that is largely an evolution of the steam engine. In other words it is not revolution it is merely evolution. This does not seem practical in our world that contains so many things that are electronic and are of a revolutionary nature. It is true that our automobiles today contain many things of an innovative measure such as GPS radar etc. however the basic engine is practically the same as it was many years ago. A rotary engine that does not reverse momentum could offer such a tremendous advantage in a time when fuel is so expensive.

This is a cutaway section of one of 2 master sections each containing three micro cylinders. These sections are geared together. So that gasses from one section feed into another. Gases are exchanged between the three micro cylinders and the two master cylinders are geared to the main shaft. Each micro cylinder is 4 inches deep.

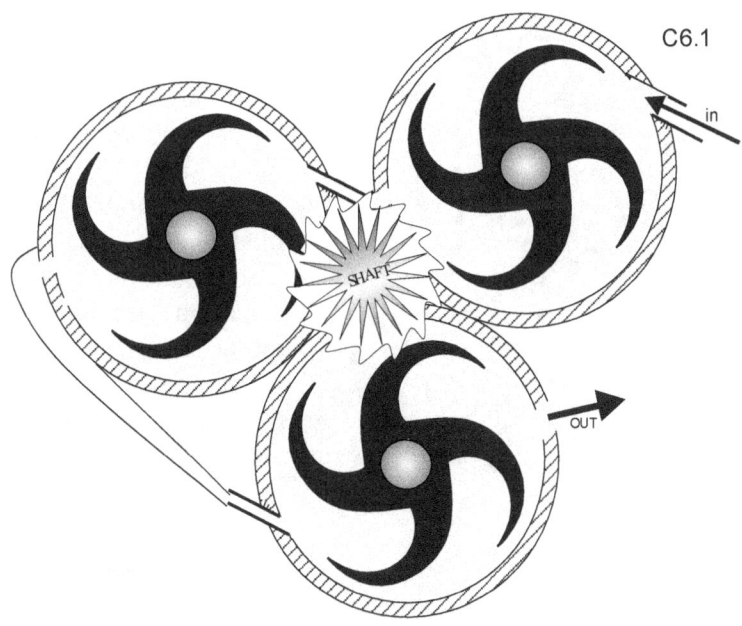

C6.1

The micro cylinder below is driven by one side that captures the gas and the other side that allows the gasses to flow freely past. This causes shaft rotation. The shaft is geared to 2 other micro shafts that connect to the master shaft. They all work so that no momentum is reversed, which will result in much more power output and faster rotational start up.

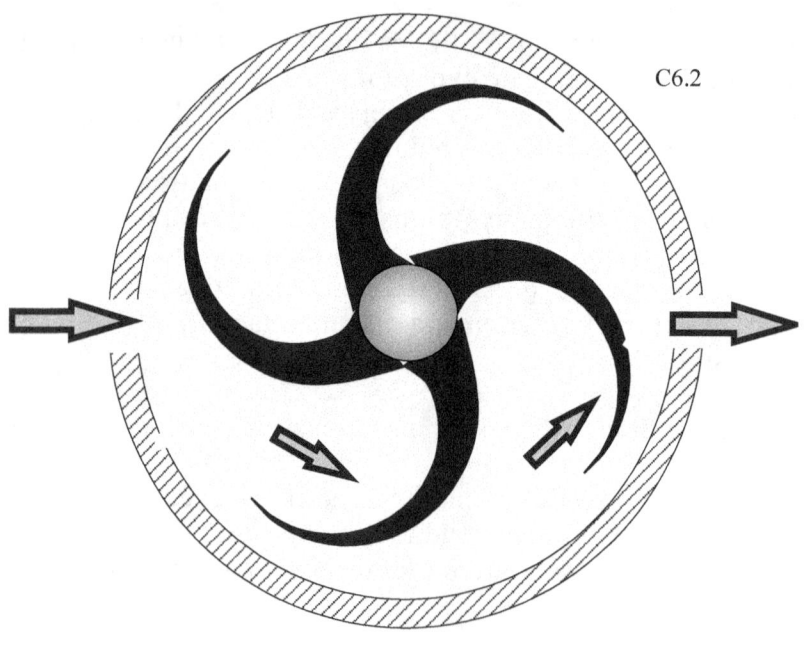

C6.2

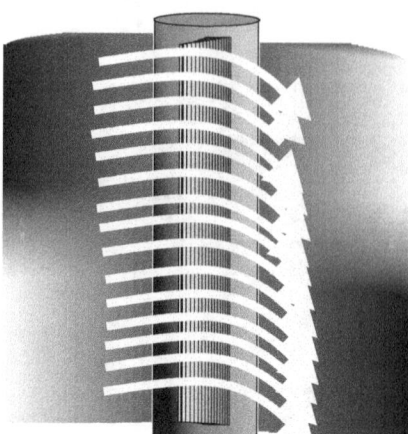

The gases flow over the top of the micro cylinder blade and flow into curled portion of the other blade. This gives aerodynamics in one direction and in the other direction a great amount of resistance. It is this resistance that allows the shaft motion

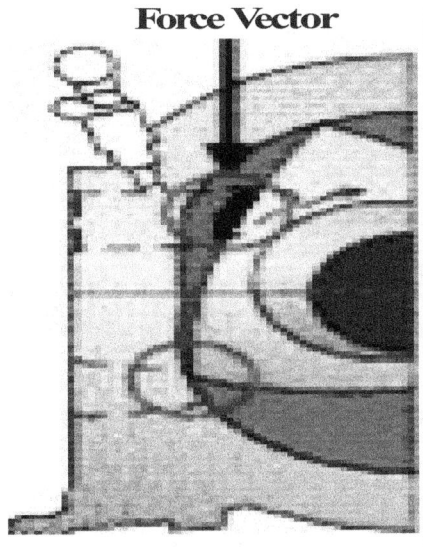

Force Vector

To the left is a side cut of the only Rotary engine in use today. It should be noted that the force factor is skewed with the shaft and therefore much of the power is lost.

Our engine does not use this principle. All force vectors are 90° in alignment with the shaft. This is but one advantage of our concept of invention. In fact this is such an erroneous problem for the general Rotary engine that its elimination would practically guarantee a superior engine to the internal combustion engine.

"Even modern motor vehicles with fossil fueled internal combustion engines are relatively inefficient, especially compared with other forms of transportation such as railroad trains when payload-miles/gallon are considered. U.S. Department of Energy data indicates that an average of 12.6 percent of the energy as a result of combustion in a gasoline engine actually makes it all the way to the rear wheels. However, after engine losses, horsepower at the flywheel is considered to lose about 20 to 22 percent of its value on its way to the rear wheels through the transmission, driveshaft and rear axle."[1]

---

[1] http://www.lightrailnow.org/facts/fa_lrt_2007-08a.htm

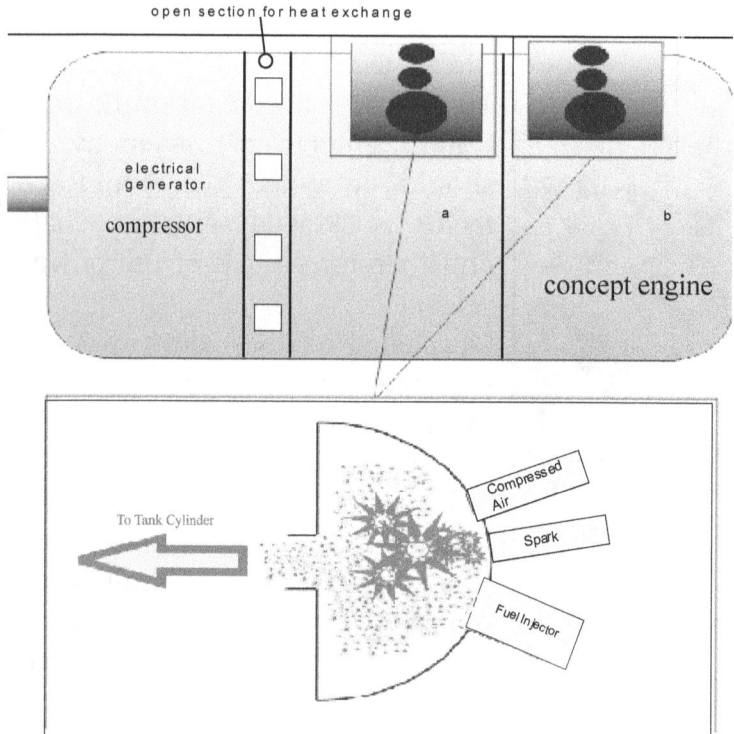

open section for heat exchange

electrical generator

compressor

a

b

concept engine

To Tank Cylinder

Compressed Air

Spark

Fuel Injector

Each micro chamber has one combustion chamber. The combustion chamber is fed by compressed air, fuel injection, and a spark. This mixture is ignited in the combustion chamber as a compressed explosive force which propels the shaft which in turn propels the air compressor.

Upon initial ignition there is little resistance from the shaft because there is no internal reversal of the pistons and other shaft parts. This means that a high velocity starter can be used along with a lithium battery instead of a extremely high current Edison cell. **This also allows the compressor** to offer compressed air in a short period time. This Rotary engine offers a high velocity angular acceleration.

Section C7. **Retractable studs**

There is a way to make retractable studs. If there is a way to measure tire pressure through wireless(and there is) then there should be a way to control a valve inside of your tires to have studs by remote control. One way to do this will be shown there may of course be other ways.

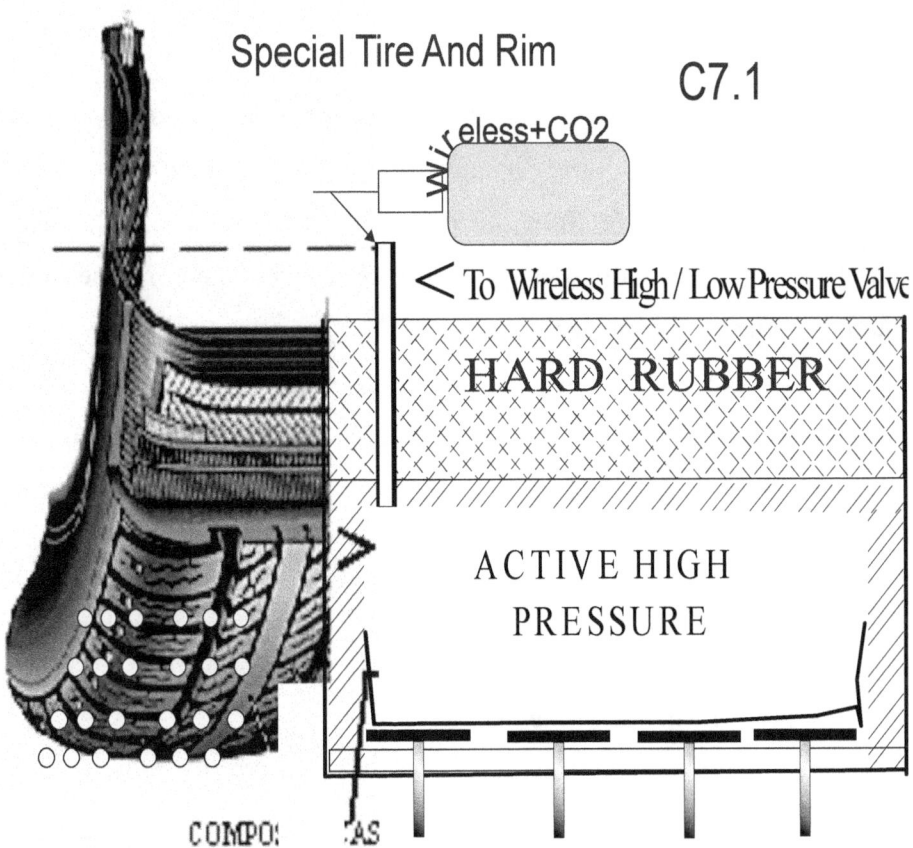

Special Tire And Rim

C7.1

Wireless+CO2

< To Wireless High / Low Pressure Valve

HARD RUBBER

ACTIVE HIGH PRESSURE

COMPO: AS

It is important to remember that with the advent of wireless tire pressure measurement, power is being generated inside the wheel. If it were not for this power wireless studs would not be possible.

The power inside the wheel is used to activate wireless receivers that are controlled from inside the car. The receivers are contained inside of each wheel, but the control transmitters are allowed to active only the back or front tires at one time.

It can be seen from figure C7.1 that studs have been activated. A wireless signal is sent from the control panel inside the car. A slow turn worm gear is used to open a valve from the co2. The co2 pressurizes the inside chamber of the specially designed tire. A rubber sheet in the inside chamber of the tire becomes pressurized.

The rubber sheet in the inside chamber coverers the top of all of the specially designed studs. The studs are designed so that one side is fixed to a disc and the other side pushes down thrown through the holes in the tire.

If the tire is deactivated by the wireless the co2 will leave the inside chamber of the tire and close a valve from the co2. And the stud will be pushed back to the inside of the tire.

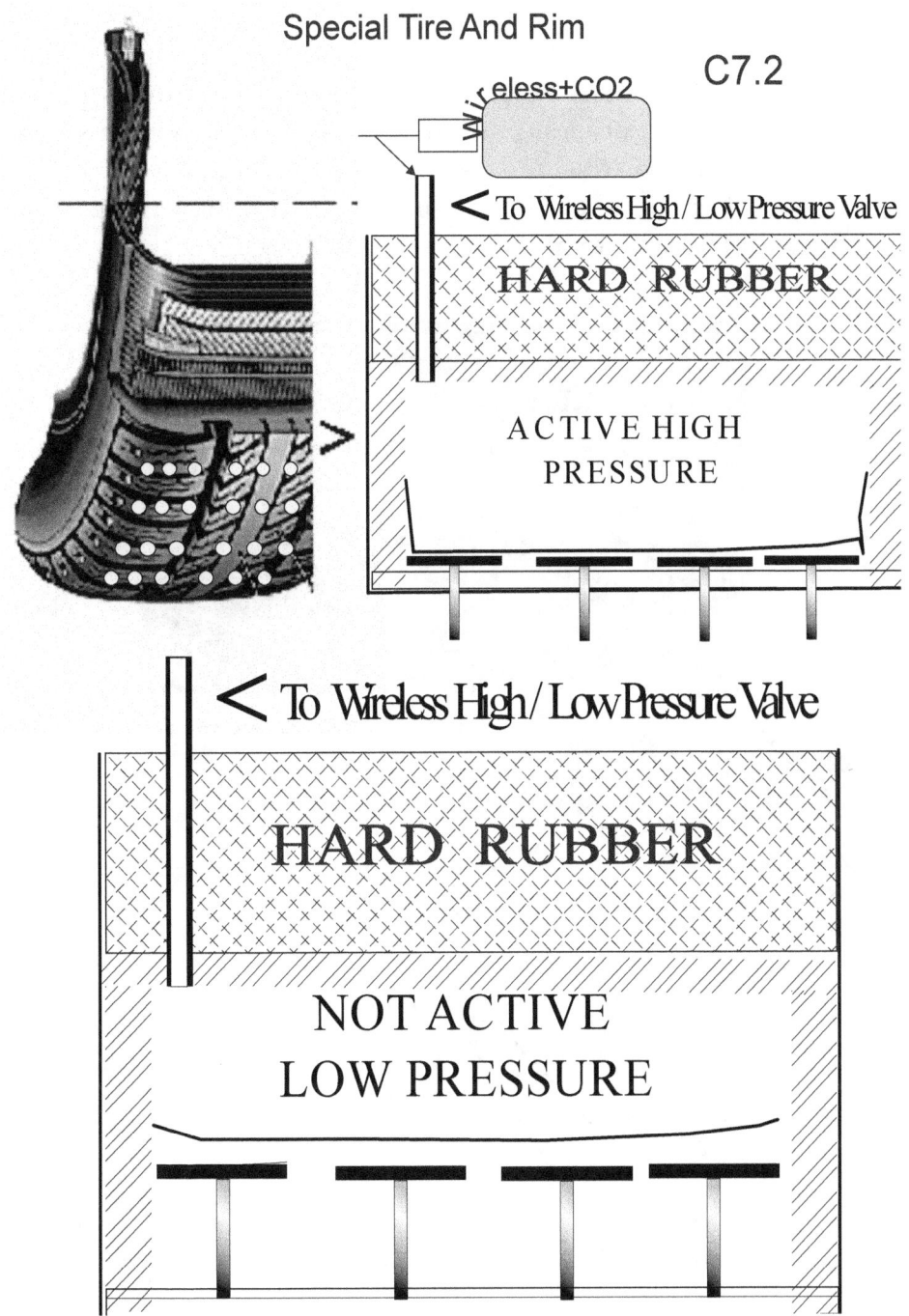

Special Tire And Rim

C7.2

Wireless+CO2

< To Wireless High / Low Pressure Valve

HARD RUBBER

ACTIVE HIGH
PRESSURE

< To Wireless High / Low Pressure Valve

HARD RUBBER

NOT ACTIVE
LOW PRESSURE

Section C8.Solar House

The solar house below is probably the most common type and resides in a warm are hot portion of the United States. This type of configuration is relatively useless in   most of the United States.
A much better concept of invention can be utilized from this ordinary passive system.

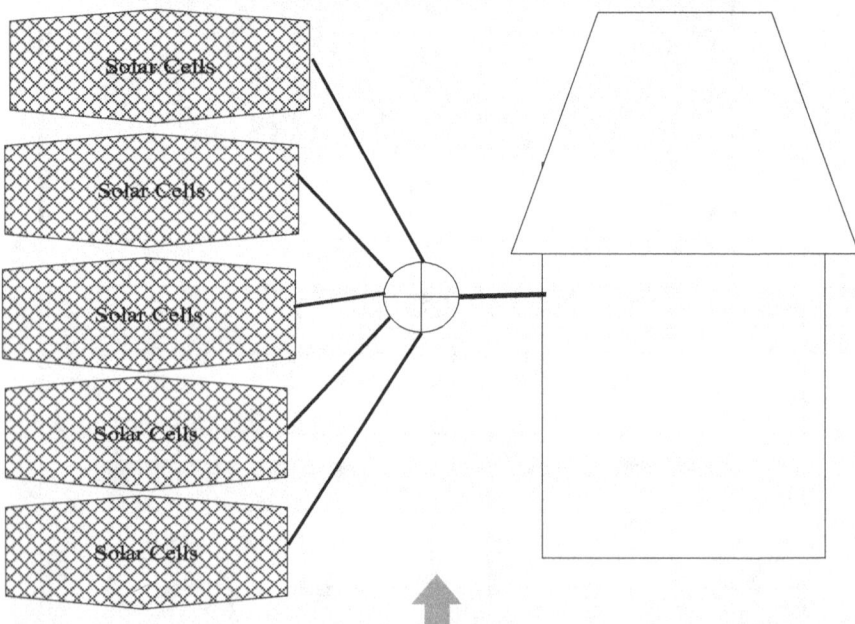

This is a passive system with long-term battery backup in which an intelligent measure of electricity can be used. In occasions he electricity can be stored overnight. Especially if it is conserved during the day and the house is located in a certain part of the United States.

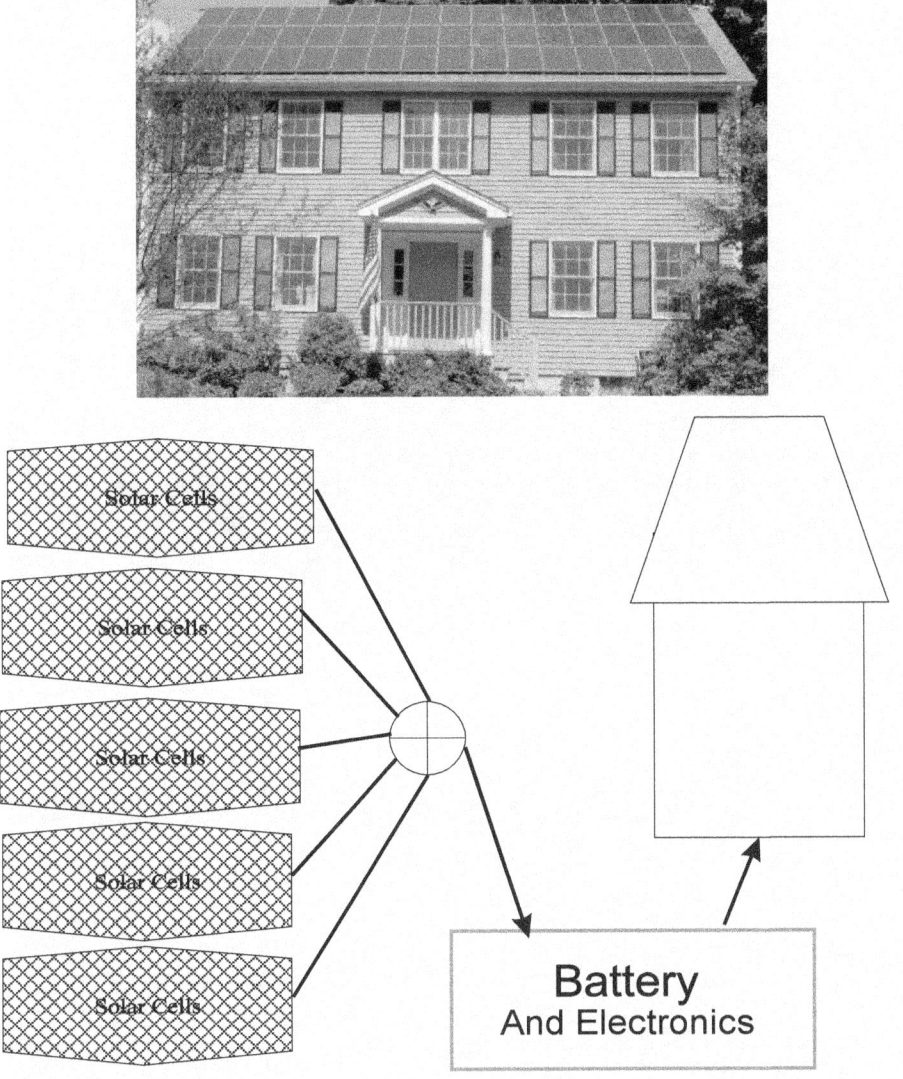

This system uses backup Rotary storage that can slowly charge the battery. The storage system allows the battery to reach full charge and in many parts of the United States this system would provide full coverage.

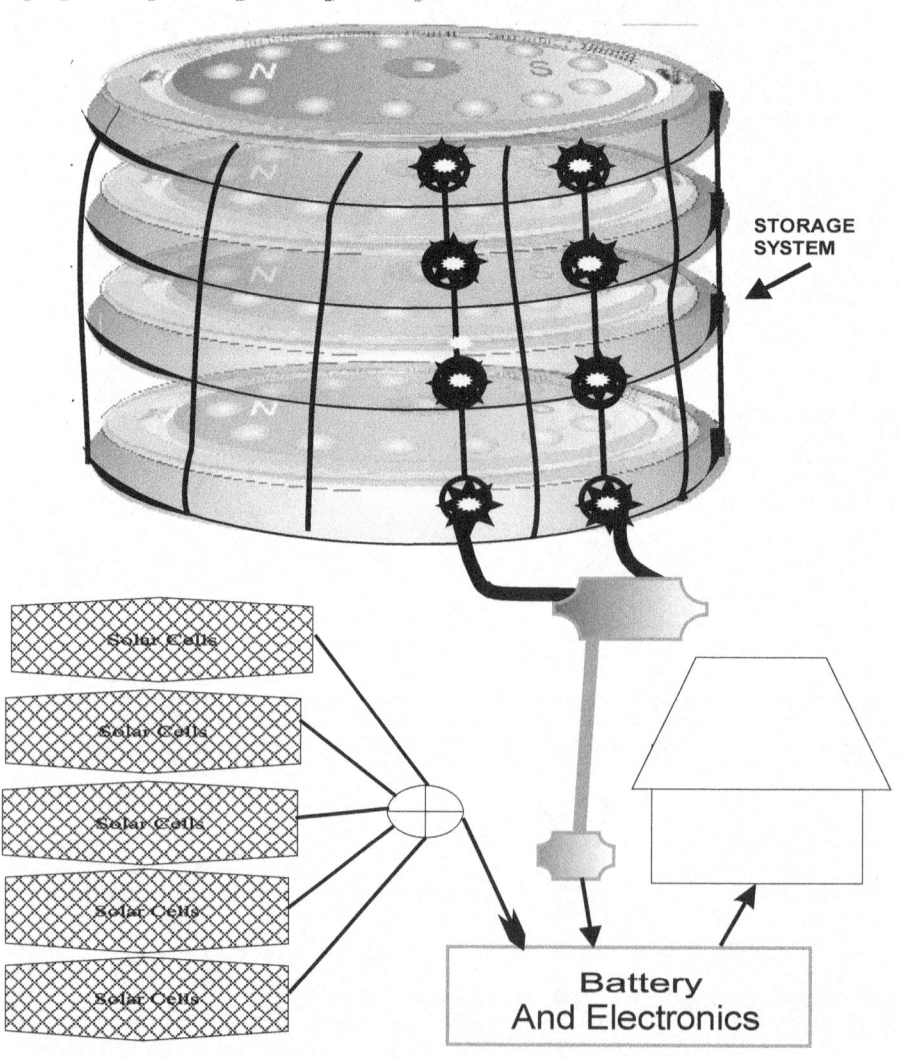

STORAGE
SYSTEM

Solar Cells

Solar Cells

Solar Cells

Solar Cells

Solar Cells

Battery
And Electronics

"**BERKELEY, CA** ◆ Researchers in the Materials Sciences Division (MSD) of Lawrence Berkeley National Laboratory, working with crystal-growing teams at Cornell University and Japan's Ritsumeikan University, have learned that the band gap of the semiconductor indium nitride is not 2 electron volts (2 eV) as previously thought, but instead is a much lower 0.7 eV.

The serendipitous discovery means that a single system of alloys incorporating indium, gallium, and nitrogen can convert virtually the full spectrum of sunlight -- from the near infrared to the far ultraviolet -- to electrical current.

"It's as if nature designed this material on purpose to match the solar spectrum," says MSD's Wladek Walukiewicz, who led the collaborators in making the discovery.

What began as a basic research question points to a potential practical application of great value. For if solar cells can be made with this alloy, they promise to be rugged, relatively inexpensive -- and most efficient. Many factors limit the efficiency of photovoltaic cells. Silicon is cheap, for example, but in converting light to electricity it wastes most of the energy as heat. The most efficient semiconductors in solar cells are alloys made from elements from group III of the periodic table, like aluminum, gallium, and indium, with elements from group V, like nitrogen and arsenic.

One of the most fundamental limitations on solar cell efficiency is the band gap of the semiconductor from which the cell is made. In a photovoltaic cell, negatively doped (n-type) material, with extra electrons in its otherwise empty conduction band, makes a junction with positively doped (p-type) material, with extra holes in the band otherwise filled with valence electrons. Incoming photons of the right energy -- that is, the right color of light -- knock electrons loose and leave holes; both migrate in the junction's electric field to form a current.

Photons with less energy than the band gap slip right through. For example, red light photons are not absorbed by high-band-gap semiconductors. While photons with energy higher than the band gap are absorbed -- for example, blue light photons in a low-band gap semiconductor -- their excess energy is wasted as heat.

The maximum efficiency a solar cell made from a single material can achieve in converting light to electrical power is about 30 percent; the best efficiency actually achieved is about 25 percent. To do better, researchers and manufacturers stack different band gap materials in multijunction cells.

Dozens of different layers could be stacked to catch photons at all energies, reaching efficiencies better than 70 percent, but too many problems intervene. When crystal lattices differ too much, for example, strain damages the crystals. The most efficient multijunction solar cell yet made -- 30 percent, out of a possible 50 percent efficiency -- has just two layers."[1]

[1] http://www.lbl.gov/Science-Articles/Archive/MSD-full-spectrum-solar-cell.html

Section C9.Ethanol Still

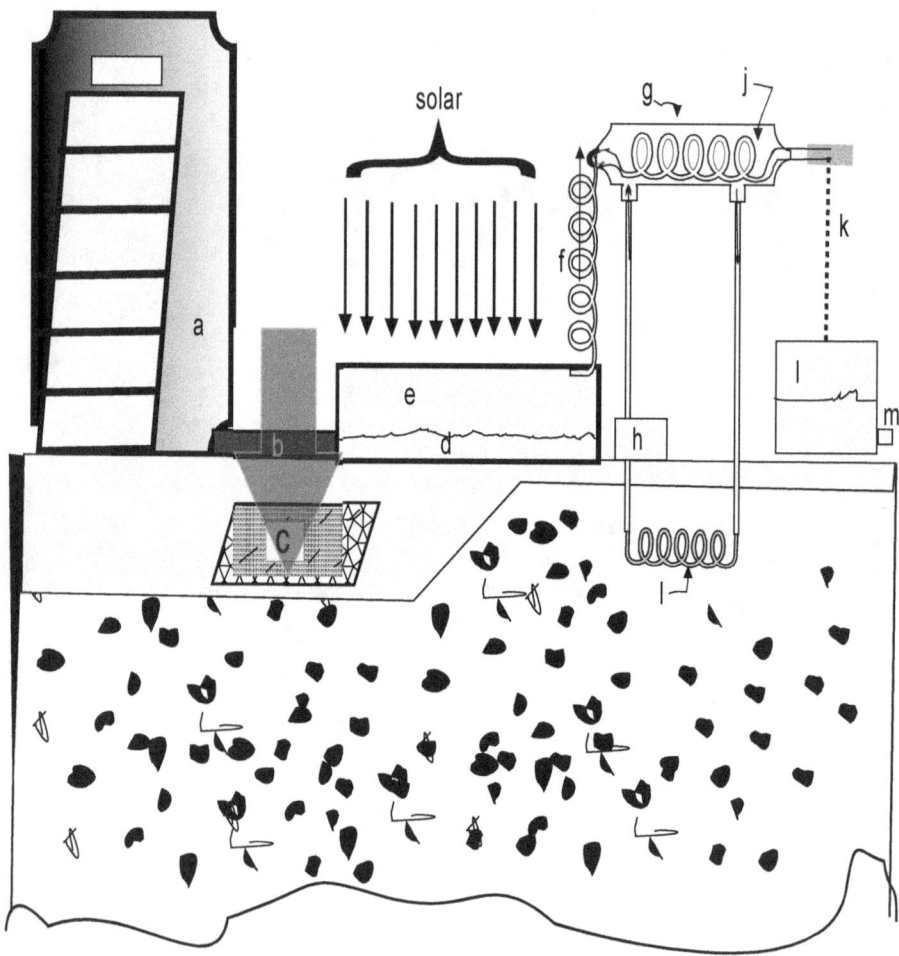

A Mixture container
B Solar
C Solar cell
D Mixture
E Mixture vapor
F Fall back tube
G Condenser shield
H Coolant pump
I Coolant
J Condenser tubing
K   Refined solution
L Container
M Sprout

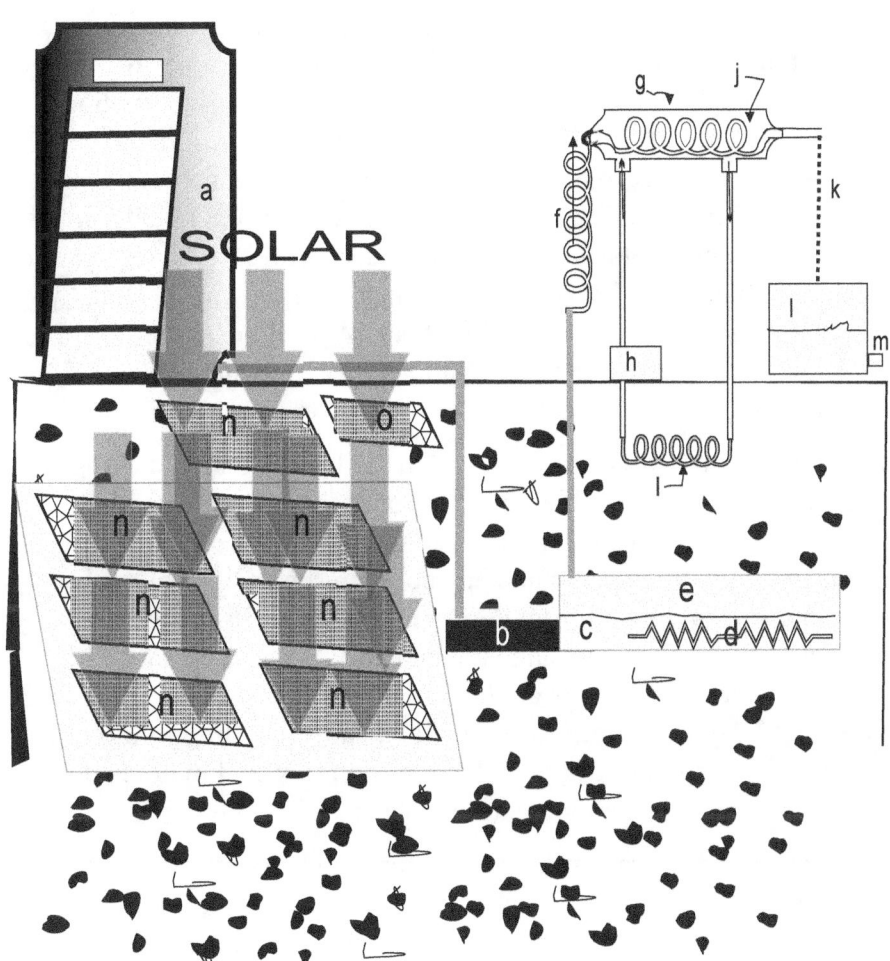

A Mixture container
B Low flow tank
C High flow tank
D Heating element
E Mixture vapor
F Fall back tube
G Condenser shield
H Coolant pump
I Refined solution
J Condenser tubing
K Refined solution
L Container
M Sprout
N Solar cells
O Solar cell

"**Ethanol fuel** is ethanol (ethyl alcohol), the same type of alcohol found in alcoholic beverages. It is most often used as a motor fuel, mainly as a biofuel additive for gasoline. World ethanol production for transport fuel tripled between 2000 and 2007 from 17 billion to more than 52 billion liters. From 2007 to 2008, the share of ethanol in global gasoline type fuel use increased from 3.7% to 5.4% In 2011 worldwide ethanol fuel production reached 22.36 billionU.S. liquid gallons (bg) (84.6 billion liters), with the United States as the top producer with 13.9 bg (52.6 billion liters), accounting for 62.2% of global production, followed by Brazil with 5.6 bg (21.1 billion liters).[2] Ethanol fuel has a "gasoline gallon equivalency" (GGE) value of 1.5 US gallons (5.7 L), which means 1.5 gallons of ethanol produce the energy of one gallon of gasoline

Ethanol fuel is widely used in Brazil and in the United States, and together both countries were responsible for 87.1% of the world's ethanol fuel production in 2011 Most cars on the road today in the U.S. can run on blends of up to 10% ethanol and ethanol represented 10% of the U.S. gasoline fuel supply derived from domestic sources in 2011 Since 1976 the Brazilian government has made it mandatory to blend ethanol with gasoline, and since 2007 the legal blend is around 25% ethanol and 75% gasoline (E25) By December 2011 Brazil had a fleet of 14.8 million flex-fuel automobiles and light trucks and 1.5 million flex-fuel motorcycles that regularly use neat ethanol fuel (known as E100).

Bioethanol is a form of renewable energy that can be produced from agricultural feeds tocks. It can be made from very common crops such as sugar cane, potato, manioc and corn. There has been considerable debate about how useful bioethanol will be in replacing gasoline. Concerns about its production and use relate to increased food prices due to the large amount of arable land required for crops, as well as the energy and pollution balance of the whole cycle of ethanol production, especially from corn. Recent developments with cellulosic ethanol production and commercialization may allay some of these concerns."[2]

"Ethanol fuel is ethanol (ethyl alcohol). Ethanol, also called ethyl alcohol, pure alcohol, grain alcohol, or drinking alcohol, is a volatile, flammable, colorless liquid. A psychoactive drug and one of the

---

[2] http://en.wikipedia.org/wiki/Ethanol_fuel

oldestrecreational drugs known, ethyl alcohol produces a state known as alcohol intoxication when consumed. Best known as the type of alcohol found in alcoholic beverages, it is also used in thermometers, as a solvent, and as a fuel. In common usage, it is often referred to simply as alcohol or spirits. the same type of alcohol found in alcoholic beverages.

It is most often used as a motor fuel, mainly as a biofuel additive for gasoline. World ethanol production for transport fuel tripled between 2000 and 2007 from 17 billion to more than 52 billion litres. From 2007 to 2008, the share of ethanol in global gasoline type fuel use increased from 3.7% to 5.4%. In 2011 worldwide ethanol fuel production reached 22.36 billion U.S. liquid gallons (bg) (84.6 billion liters), with the United States as the top producer with 13.9 bg (52.6 billion liters), accounting for 62.2% of global production, followed by Brazil with 5.6 bg (21.1 billion liters). Ethanol fuel has a "gasoline gallon equivalency" (GGE) value of 1.5 US gallons (5.7 L), which means 1.5 gallons of ethanol produce the energy of one gallon of gasoline.

Ethanol fuel is widely used in Brazil and in the United States, and together both countries were responsible for 87.1% of the world's ethanol fuel production in 2011. Most cars on the road today in the U.S. can run on blends of up to 10% ethanol, and ethanol represented 10% of the U.S. gasoline fuel supply in 2011. Since 1976 the Brazilian government has made it mandatory to blend ethanol with gasoline, and since 2007 the legal blend is around 25% ethanol and 75% gasoline (E25). By December 2011 Brazil had a fleet of 14.8 million flex-fuel automobiles and light trucksand 1.5 million flex-fuelmotorcycles that regularly use neat ethanol fuel (known as E100)." [3]

---

[3] http://www.ourmed.org/wiki/Ethanol

Section C10.Safe Nuclear Battery

There is at present no difficulty in extracting safe energy from the atom but a much bigger problem is the stigma that has been attached to nuclear energy. This problem may be more mental than actual.

If you saw a snake in the woods and didn't know what kind of snake it was you would be very concerned because you wouldn't know if it was poisonous are not poisonous. The same thing is true with radiation if you don't know what type of radiation is harmful you could assume that all types of radiation are harmful and you would not apply it to any purpose even though it may be very useful.

Fusion is a much safer form of energy but has never been realized to date. If fusion energy were available it would be a very safe form of wave energy with no byproducts. Of course the problem with obtaining fusion is in obtaining the amount of compression necessary for the process to make any reaction. It is unlikely that fusion will be obtained. Fission is very unsafe, but at present is the only one we are using. There are many safeguards that must be in place for the implementation of this primitive form of energy extraction. There ia a psychological fear of any of the terms evolving from the word nuclear radiation however there are ways to use the byproducts of fission to make batteries or generators that can last a very long time these are commonly referred to as radio isotopic batteries, but sometimes nuclear batteries.

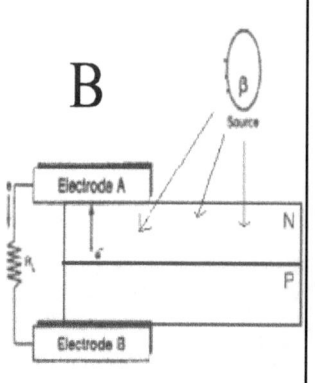

In the figure on the left a PN junction is being bombarded by beta particles. It should be noted that the bombardment is not inside the junction but the beta particles must pass through the junction in order to induce electron flow. In concepts of inventions a better battery will be produced by including the beta particles within the PN junction.

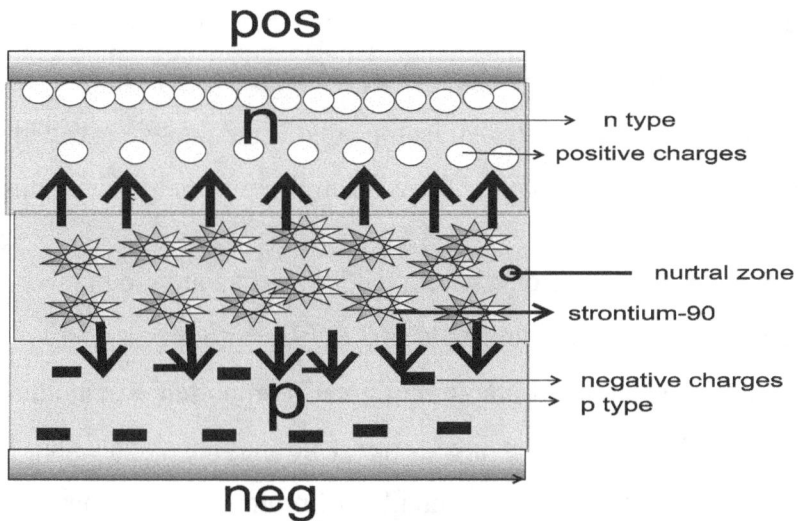

There are only two types of particle relative radioactive materials that are relatively safe. These are the beta particles and alpha particles; however the alpha particles have a tendency to degrade the semiconductor materials and shall not be used. There are only a few radioactive isotopes that emit beta particles we will use Strontium-90 which does not exist in nature, but it is one of the several radioactive waste products resulting from nuclear fission. The utilizable energy from strontium-90 substantially exceeds the energy derived from the nuclear fission which gave rise to this isotope.

Particles of the strontium 90 are inserted in the PN junction for maximum effect. Given that the strontium 90 as a half-life of about 40 years this provides a very effective battery.

"What are isotopes?

The isotopes of an element are all the atoms that have in their nucleus the number of protons (atomic number) corresponding to the chemical behavior of that element. However, the isotopes of a single element vary in the number of neutrons in their nuclei. Since they still have the same number of protons, all these isotopes of an element have *identical chemical behavior*. But since they have different numbers of neutrons, these isotopes of the same element may have *different radioactivity*. An isotope that is radioactive is called a *radioisotope* or *radionuclide*. Two examples may help clarify this.

The most stable isotope of uranium, U-238, has an atomic number of 92 (protons) and an atomic weight of 238 (92 protons plus 146 neutrons). The isotope of uranium of greatest importance in atomic bombs, U-235, though, has three fewer neutrons. Thus, it also has an atomic number of 92 (since the number of protons has not changed) but an atomic weight of 235 (92 protons plus only 143 neutrons). The *chemical behavior* of U-235 is identical to all other forms of uranium, but its nucleus is less stable, giving it *higher radioactivity* and greater susceptibility to the chain reactions that power both atomic bombs and nuclear fission reactors.

Another example is iodine, an element essential for health; insufficient iodine in one's diet can lead to a goiter. Iodine also is one of the earliest elements whose radioisotopes were used in what is now called nuclear medicine. The most common, stable form of iodine has an *atomic number* of 53 (protons) and an *atomic weight* of 127 (53 protons plus 74 neutrons). Because its nucleus has the "correct" number of neutrons, it is stable and is not radioactive. A less stable form of iodine also has 53 protons (this is what makes it behave chemically as iodine) but four extra neutrons, for a total atomic weight of 131 (53 protons and 78 neutrons). With "too many" neutrons in its nucleus, it is unstable and radioactive, with a half-life of eight days.

Because it behaves *chemically* as iodine, it travels throughout the body and localizes in the thyroid gland just like the stable form of iodine. But, because it is radioactive, its presence can be detected. Iodine 131 thus became one of the earliest *radioactive tracers.*

How can different isotopes of an element be produced?

How can isotopes be produced--especially radioisotopes, which can serve many useful purposes? There are two basic methods: separation and synthesis.

Some isotopes occur in nature. If radioactive, these usually are radioisotopes with very long half-lives. Uranium 235, for example, makes up about 0.7 percent of the naturally occurring uranium on the earth The challenge is to *separate* this very small amount from the much larger bulk of other forms of uranium. The difficulty is that all these forms of uranium, because they all have the same number of electrons, will have identical chemical behavior: they will bind in identical fashion to other atoms. Chemical separation, developing a chemical reaction that will bind only uranium atoms, will separate out uranium atoms, but not distinguish among different isotopes of uranium. The only difference among the uranium isotopes is their *atomic weight.* A method had to be developed that would sort atoms according to weight.

One initial proposal was to use a *centrifuge.* The basic idea is simple: spin the uranium atoms as if they were on a very fast merry-go-round. The heavier ones will drift toward the outside faster and can be drawn off. In practice the technique was an enormous challenge: the goal was to draw off that very small portion of uranium atoms that were lighter than their brethren. The difficulties were so enormous the plan was abandoned in 1942. Instead, the technique of *gaseous diffusion* was developed. Again, the basic idea was very simple: the rate at which gas passed (*diffused*) through a filter depended on the weight of the gas molecules: lighter molecules diffused more quickly. Gas molecules that contained U-235 would diffuse slightly faster than gas molecules containing the more common but also heavier U-238. This method also presented formidable technical challenges, but was eventually implemented in the gigantic gas diffusion plant at Oak Ridge, Tennessee. In this process, the uranium was chemically combined with fluorine to form a hexafluoride gas prior to separation by diffusion. This is not a practical method for extracting radioisotopes for scientific and medical use. It was extremely expensive and could only supply naturally occurring isotopes.

A more efficient approach is to artificially manufacture radioisotopes. This can be done by firing high-speed particles into the nucleus of an atom. When struck, the nucleus may absorb the particle or become unstable and emit a particle. In either case, the number of particles in the nucleus would be altered, creating an isotope. One source of high-speed particles could be a cyclotron. A cyclotron accelerates particles around a circular race track with periodic pushes of an electric field. The particles gather speed with each push, just as a child swings higher with each push on a swing. When traveling fast enough, the particles are directed off the race track and into the target.

A cyclotron works only with charged particles, however. Another source of bullets are the neutrons already shooting about inside a nuclear reactor. The neutrons normally strike the nuclei of the fuel, making them unstable and causing the nuclei to split (fission) into two large fragments and two to three "free" neutrons. These free neutrons in turn make additional nuclei unstable, causing further fission. The result is a chain reaction. Too many neutrons can lead to an uncontrolled chain reaction, releasing too much heat and perhaps causing a "meltdown." Therefore, "surplus" neutrons are usually absorbed by "control rods." However, these surplus neutrons can also be absorbed by targets of carefully selected material placed in the reactor. In this way the surplus neutrons are used to create radioactive isotopes of the materials placed in the targets.

With practice, scientists using both cyclotrons and reactors have learned the proper mix of target atoms and shooting particles to "cook up" a wide variety of useful radioisotopes."[4]

"Any radioisotope in the form of a solid that gives off alpha or beta particles can be utilized in the nuclear battery. The first cell constructed (that melted the wire components) employed the most powerful source known, radium-226,as the energy source. However, radium 226 gives rise through decay to the daughter product bismuth-214, which gives off strong gamma radiation that requires shielding for safety. This adds a weight penalty in mobile applications.

---

[4] http://en.wikipedia.org/wiki/Isotope

Strontium-90 gives off no gamma radiation so it does not necessitate the use of thick lead shielding for safety. Strontium-90 does not exist in nature, but it is one of the several radioactive waste products resulting from nuclear fission. The utilizable energy from strontium-90 substantially exceeds the energy derived from the nuclear fission which gave rise to this isotope.

l is several times than conventional energy sources, and is by far the highest, the cells are comparatively much lighter and thus facilitates high energy densities to be achieved. Similarly, the efficiency of such cells is much higher simply because radioactive materials involve little waste generation. Thus substituting the future energy needs with nuclear cells and replacing the already existing ones with these, the world can be seen transformed into a new one by reducing the green house effects and associated risks. This should come as a handy savior for almost all developed and developing nations worldwide. More over the nuclear waste produced there in are substances that don't occur naturally. For example, Strontium-90 does not exist in nature, but it is one of the several radioactive waste products resulting from nuclear fission."[5]

"The Elgin-Kidde nuclear battery is said to be immune to heat, cold, and self- damage by radiation. Short-circuiting it does no harm, and power drawn from it ...

---

[5]http://worldwidescience.org/topicpages/s/source+radioisotope+thermoelectric.html

Some have dubbed it a "nuclear battery" since it will run without refueling for its entire 30-year lifetime. The key to the hands-off maintenance plan for ...

Atomic batteries use the heat given off by the natural decay of various radioactive isotopes, such as plutonium 238, curium 244, or promethium .

All of these plants will be nuclear-powered, dual-purpose ones, producing vast .... and raclioisotopes of the kind used in atomic batteries for spacecraft.

Simple, efficient atomic batteries that convert nuclear energy directly to electricity are being developed by University of Illinois researchers GH Miley ...

The third one uses a SNAP-9A nuclear-power source— an atomic battery in which heat from plutonium 238 is converted by thermocouples into about 20 watts of ...

Your article, "Electric Car Showdown in Phoenix: Zinc-Air Battery Wins" [July] ... earthbound means of generating non-polluting power — nuclear plants.

Nuclear energy exists in the far stronger, short-range forces that hold the .... Located near the detonators are thermal batteries, devices that have a long ...>"[6789]

---

[6]

https://www.ideals.illinois.edu/bitstream/handle/2142/16849/1_Yakubova_Galina.pdf?sequence=3

[7]

http://books.google.com/books?id=sP7ag1e4tzEC&pg=PA9&lpg=PA9&dq=Your+article,+%22Electric+Car+Showdown+in+Phoenix:+Zinc-Air+Battery+Wins%22+%5BJuly%5D+...+earthbound+means+of+generating+non-polluting+power+%E2%80%94+nuclear+plants.&source=bl&ots=9gSWuvqqcQ&sig=wlpvlFA2S9n2O3gYDa5pioTlTio&hl=en&sa=X&ei=SaKvUMipFo6kqQGBp4HoDQ&ved=0CCwQ6AEwAA#v=onepage&q=Your%20article%2C%20%22Electric%20Car%20Showdown%20in%20Phoenix%3A%20Zinc-Air%20Battery%20Wins%22%20%5BJuly%5D%20...%20earthbound%20means%20of%20generating%20non-polluting%20power%20%E2%80%94%20nuclear%20plants.&f=false

---

8

http://books.google.com/books?id=9jsEAAAAMBAJ&pg=PA42&lpg=PA42&dq=Some+hav
e+dubbed+it+a+%22nuclear+battery%22+since+it+will+run+without+refueling+for+its+enti
re+30-year+lifetime.+The+key+to+the+hands-
off+maintenance+plan+for+...&source=bl&ots=m_fvwlDyYL&sig=73OFhIHjFArQqaZjHI_c
HR65mG4&hl=en&sa=X&ei=FKOvUJn0DtKCrQG5_IGgCg&ved=0CD0Q6AEwAg#v=onep
age&q=Some%20have%20dubbed%20it%20a%20%22nuclear%20battery%22%20since
%20it%20will%20run%20without%20refueling%20for%20its%20entire%2030-
year%20lifetime.%20The%20key%20to%20the%20hands-
off%20maintenance%20plan%20for%20...&f=false

9 http://en.wikipedia.org/wiki/Radioisotope_thermoelectric_generator

Section C11. **Hybrid Water Heater**

In this concept of invention a very inexpensive hybrid water heater is demonstrated. This device can be made from a regular water heater and a kit.

The core concept of the heater is that it involves a flow detector, a transistor a double pole double throw relay and a voltage doubler. The flow detector is attached to the hot water outlet from the heater. Whenever any hot water is turned on it activates the relay and inserts the voltage doubler into the circuit thus doubling the voltage and doubling the power. The circuit breaker to the hot water heater should be checked and doubled in some cases because the current will be increased by a factor of two.

Heating the hot water in line can save a large quantity of energy. Ordinarily the energy would be leaking out of a stationary hot water heater and is only needed when necessary.

In some cases a considerable amount of money can be saved by using the system. Some hybrid hot water heaters cost as much as $1600. However, almost all of the hybrid hot water heaters are over $1000.

By making these hot water heaters as a company and installing them in homes one can make a considerable business from it.

Assisted Water Heater

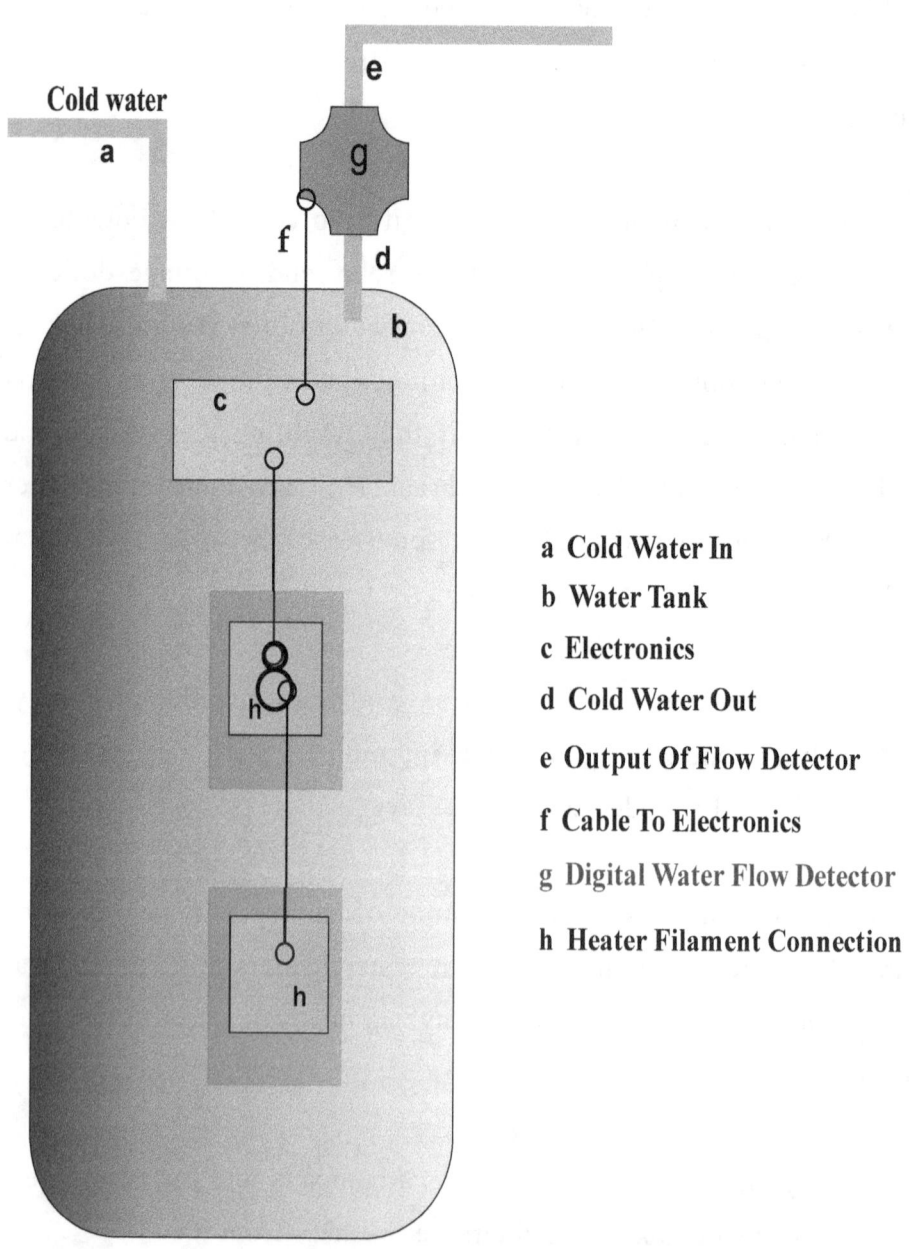

a **Cold Water In**

b **Water Tank**

c **Electronics**

d **Cold Water Out**

e **Output Of Flow Detector**

f **Cable To Electronics**

g **Digital Water Flow Detector**

h **Heater Filament Connection**

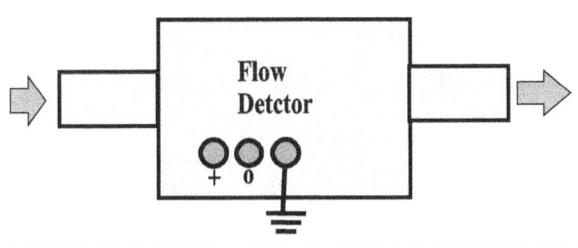

Flow Detctor

+ 0

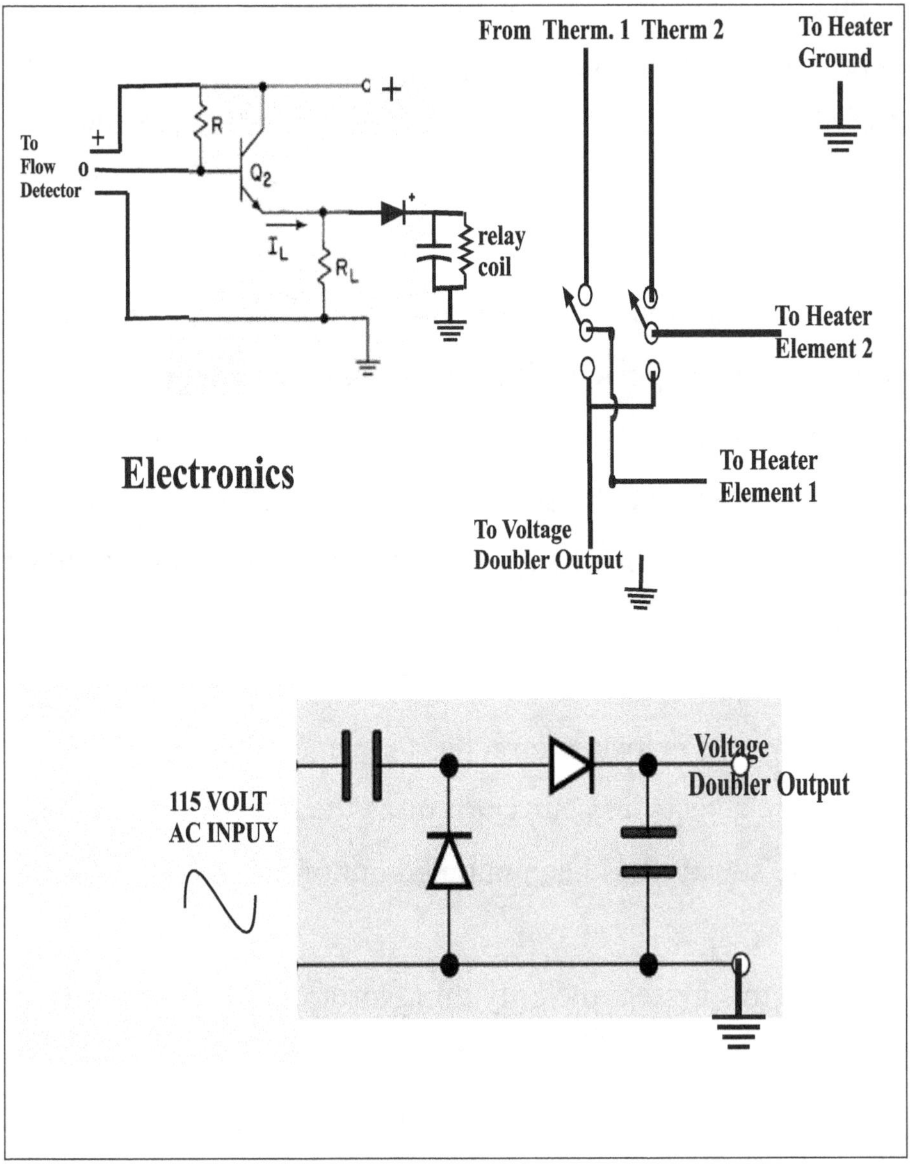

From Therm. 1 Therm 2

To Heater Ground

To Flow Detector

+ 0

R

Q_2

I_L

R_L

relay coil

To Heater Element 2

To Heater Element 1

To Voltage Doubler Output

Electronics

115 VOLT AC INPUY

Voltage Doubler Output

Section C12. Switchless Pump

This concept for invention demonstrates a switch less pump. The pump is designed specifically for sewage pumps or a sewer pump. The problem with these pumps is that the switch wears out very quickly and then a lot of money is paid to have the switch replaced.

The workings of this system are relatively simple. When the emergency switch goes off indicating that the sewage pump is full a signal is sent to the electronics which turns on the sewage pump and it starts pumping. Since the motor does more work when it's pumping the sewage out it keeps the system locked on so to speak.

After the system is locked on it has to eventually stop. When the circuitry detects less current it turns the system off. At that time all the sewage has been pumped out of the tank.

If one uses this system the only thing left to fail is the motor which will eventually go out.
There may be a little bit of experimentation required to adjust the motor to the proper input signal.

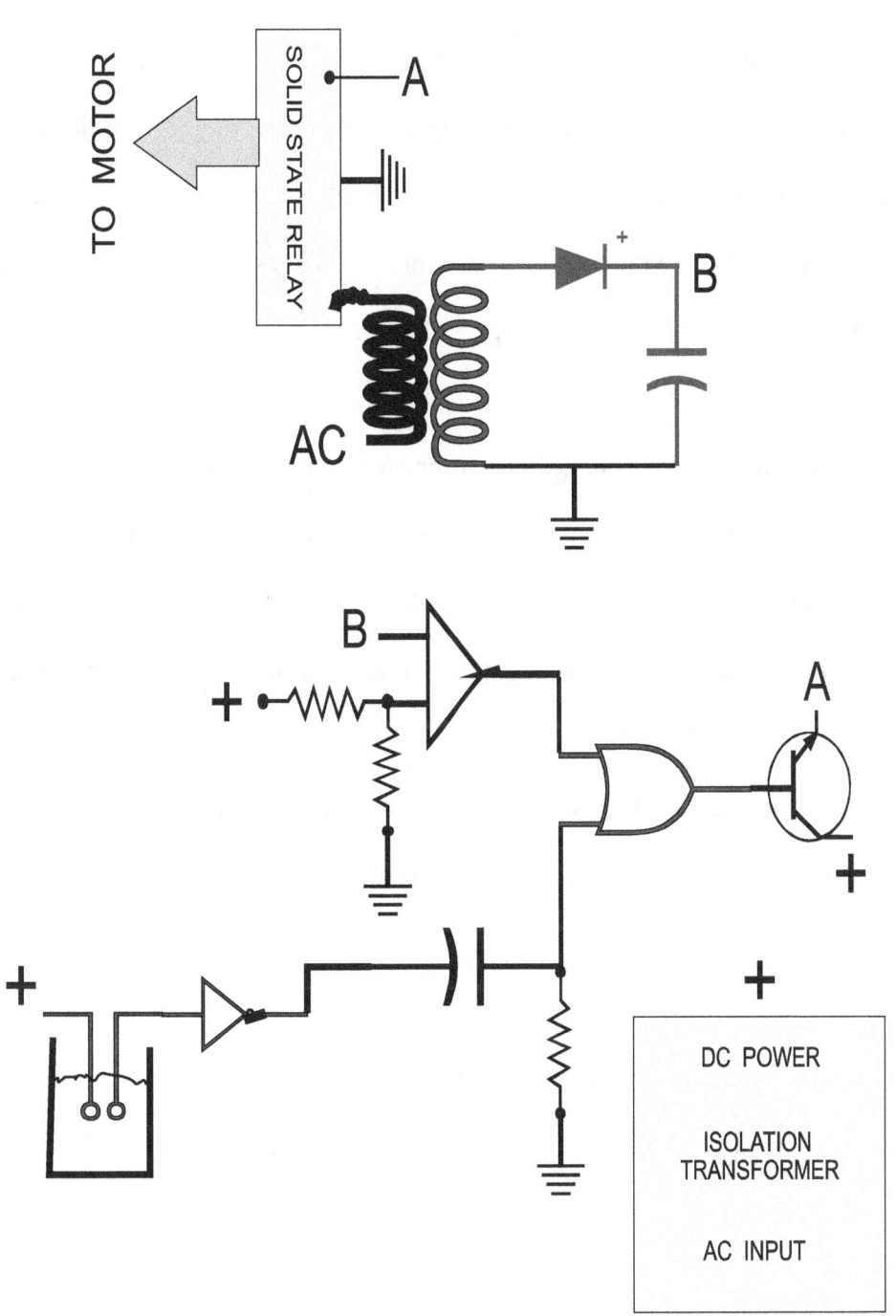

TO MOTOR

SOLID STATE RELAY

A

AC

B

DC POWER

ISOLATION
TRANSFORMER

AC INPUT

Section C13.Jet Engine Booster

The purpose of this concept of invention is to boost the velocity of a passenger jet. This is done by using some of the fuel of the jet and processing the fuel with high-efficiency electrical generators. The generators will convert the electricity into two forms a very high voltage, high current and a very low voltage and high current.

The high current low voltage is used for the magnetic field to run an MHD generator whereas the high-voltage is used to ionize the jet exhaust. Once the jet exhaust is ionized it is useful in MHD propulsion. This is expected to boost the output of the jet engines by moderate to medium amount with possibly efficient increase.

Efficient generators will be used and only a small amount of fuel will be siphoned off and converted to the energy necessary to run the MHD boosters.

Much of the MHD propulsion system is shown to be useful for experimental aircraft in the reference section.

Some of the experimental aircraft's average speeds are from Mach 5 to Mach 12.

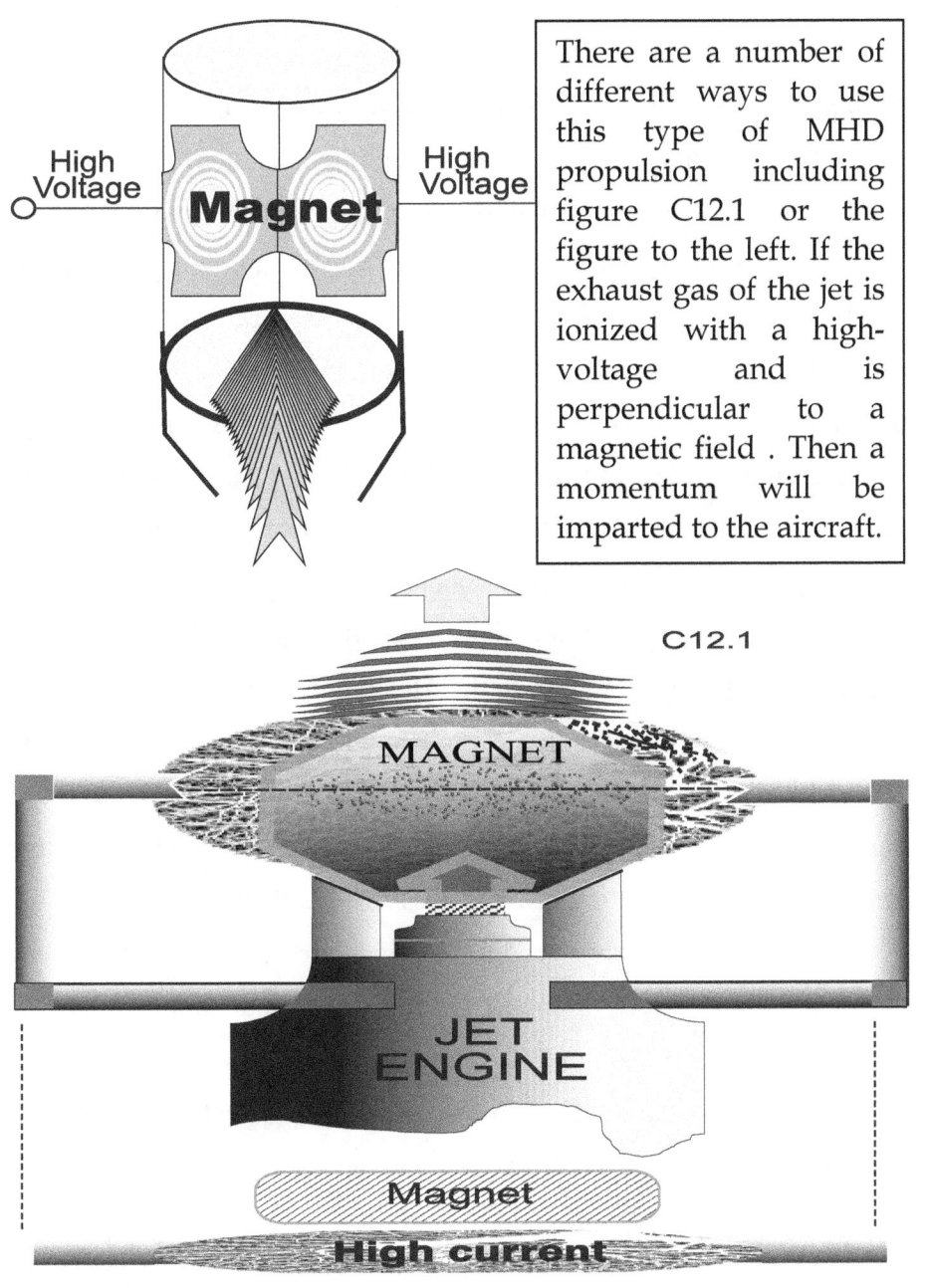

There are a number of different ways to use this type of MHD propulsion including figure C12.1 or the figure to the left. If the exhaust gas of the jet is ionized with a high-voltage and is perpendicular to a magnetic field . Then a momentum will be imparted to the aircraft.

C12.1

"A MHD (magneto-gas-dynamics, magneto-hydro-dynamics) accelerator is a device used to impart momentum to a gas through the Lorentz force. The Lorentz force is a force acting on a gas due to electromagnetic effects, and is proportional to the current and the applied magnetic field. A current can flow in a gas only if the latter is significantly ionized. Since self-ionization of air only occurs at very high temperatures that are not encountered in typical flight conditions the air is generally ionized through alkali seeding (such as potassium or cesium) or through external ionizers (such as electron beams, microwave beams or discharges for instance) to permit flow of current.One application is the magnetoplasma jet engine, which produces thrust through a MHD accelerator with the energy emanating from a stack of fuel cells. The air is ionized with electron beams as it is the most efficient way currently known to sustain ionization in cold air. Quasi-one-dimensional studies indicate that the magnetoplasma jet engine can deliver a specific impulse higher than the one of the turbojet or the ramjet in the Mach number range 3-5 while not requiring significant changes in the engine geometry. Another application of the MHD accelerator is the prevention of boundary layer separation due to shocks. Indeed, in supersonic flight, when strong shocks interact with a boundary layer, the resulting adverse pressure gradient can become high enough to cause separation. Separation regions are undesirable since they increase drag and result in more heat flow to the aircraft surfaces. Preliminary studies show that the Lorentz force created by a MHD accelerator could attenuate the adverse pressure gradient caused by the shock and hence help in preventing the separation."[1]

---

[1] http://www.bernardparent.com/viewtopic.php?f=6&t=9

70

"NASA hypersonics expert **Dr Isaiah Blankson** believes that MHD energy-conversion in the intakes can take 30-40% of the energy, letting a turbine engine run at up to Mach 7. Past the MHD the air would slow from Mach 7 to Mach 3. This was the speed of the air going into engines of the Blackbird spyplanes. The Blackbird's conventional J-58 turbojets could keep burning up to Mach 3+ because of their special intakes, which slowed the intake air down for them using a retracting central spike. This would permit the reusable first stage of a future NASA two stage to orbit launcher to take off from a runway and get its piggyback orbiter well up into scramjet-type flight regimes, all using just one set of engines."[2] "A fuel-cell powered magnetoplasma jet engine (magjet) using electron-beam ionizers is here proposed for airbreathing flight in the supersonic/hypersonic regime. The engine consists of a fuel-cell duct containing the power source and of a high-speed duct producing most of the thrust through a magnetoplasmadynamic (MHD) accelerator. To reduce the shocks and heat loads in the fuel cells, the enthalpy of the air is extracted beforehand through a MHD generator. The power produced by the latter and by the fuel cells is then split optimally between the MHD accelerator located in the high-speed duct and one located downstream of the fuel cells. The performance is assessed through exact solutions of a quasi-one-dimensional model which includes the effect of ion slip, Joule heating, and heat dissipated through electron-beam ionization. The magnetic field strength as well as the mass flow rate ratio between the high-speed and fuel cell ducts are seen to affect the thrust considerably at lower Mach number, but to have a smaller impact at hypervelocities. Flight beyond Mach 6 would necessitate substantial cooling of the fuel

---

[2] http://nextbigfuture.com/2007/12/nasa-taking-fresh-look-at-mhd.html

cells due to the ion slip effect preventing sufficient enthalpy extraction, independently of the magnetic field strength. For a fuel cell efficiency of 0.6 and a mass flow rate ratio of 5, the magjet delivers a specific impulse within 15% of the one of the turbojet in the Mach number range 1--3 given a magnetic field of 8 Teslas. From Mach 3 to 5, a magnetic field strength varying between 2 and 4 Teslas is seen to be sufficient to match the performance of conventional engiThe numerical solution of a MHD accelerator intended for supersonic airbreathing propulsion systems is presented. The numerical method solves the Favre-Averaged Navier Stokes (FANS) equations closed by the Wilcox k-omega model including the nitrogen vibrational energy and a finite-rate chemical solver accounting for electron-beam ionization, electron attachment and dissociative recombination. The fluid flow equations are solved in conjunction with the electric field potential equation. Due to the recombination time of the electrons with the charged particules being in the order of microseconds, the interaction region is more or less confined to the area when e-beam ionization is applied. In this manner, a Faraday-type configuration can be obtained by using only one electrode pair. The impact of the length of the interaction region as well as the strength of the magnetic field on the efficiency is assessed. It is observed that the efficiency obtained numerically is as much as 40% less than the theoretical predictions for the highest magnetic field considered of 4 Teslas. This is attributed to (i) the current concentration near the electrodes edges causing a significant voltage drop, and (ii) unsteady behaviour in the center of the channel"[3]

---

[3] http://www.bernardparent.com/viewtopic.php?p=21

Section C14. Fractal Stethoscope

This concept of invention allows the doctor to check your heart with a stethoscope and determine if it is in need of further examination using a new form of mathematics. If a problem is found then a computer can be attached to an EKG to determine specifically the extent of the trouble. At that point a doctor may be able to pinpoint the condition and prescribe a treatment. All of this is the result of a routine examination.

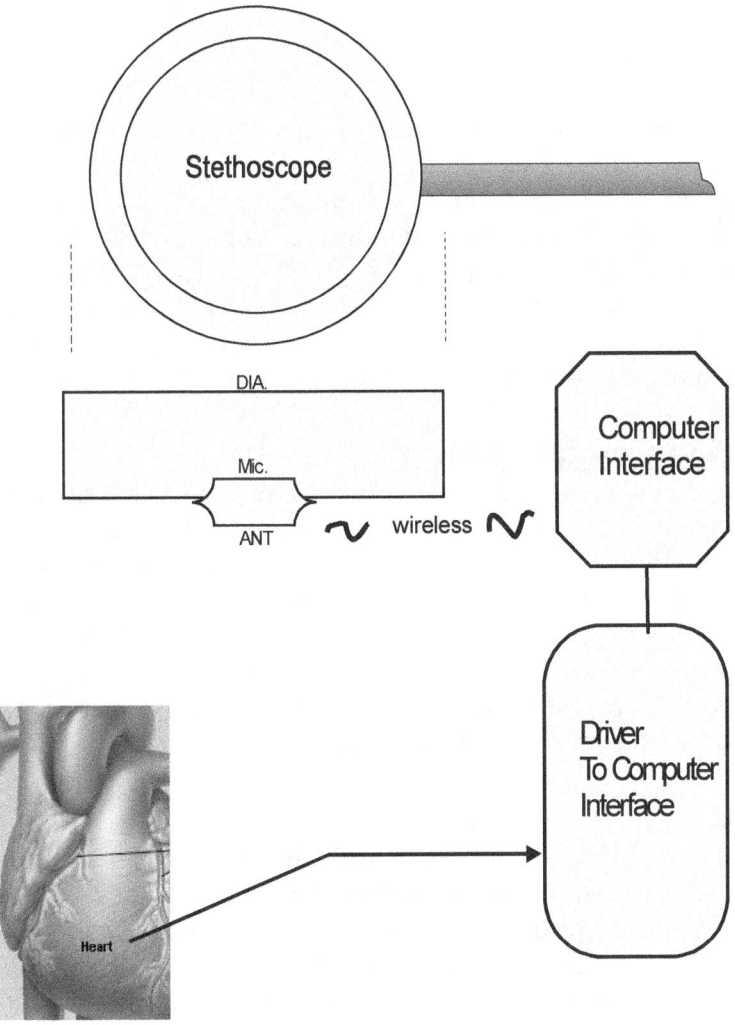

A computer can be used to analyze the heart using **Fractal Systems**.

"Fractal Systems

According to the Centre for Fractal Design and Consultancy, fractal studies is an emerging field of science that attempts to explain the movements and interactions of smaller units within a whole. Fractal studies can be used to understand neurons, atoms or molecules --- each of these units has the ability to function independently within a whole yet they also have the ability to evolve, potentially changing the behavior of that whole. By graphing and charting a unit's movement, scientists can try to identify.

Arrhythmia

Arrhythmia is a condition in which your heart doesn't beat at a consistent speed. The American Heart Association notes that although arrhythmias aren't always dangerous, they have the potential to cause your heart to pump less effectively. Sometimes an arrhythmic episode is brief; other times, they last long enough to disrupt your heart rate as measured by a doctor. This condition can be frustrating to diagnose and monitor because there is no set average heart rate against which to measure --- your heart rate changes based on your breathing rate, excitement or exertion.

Arrhythmia and Fractal Analysis

Medical scientists can apply a fractal analysis to the heartbeats of patients experiencing arrhythmia. According to L.S. Liebovitch, T. Penzel and J.W. Kantelhardt in "The Science of Disasters," fractal analysis has shown there is a pattern between an arrhythmic patient's periods of rapid heart rate and the patient's periods of additional heartbeats. Although fractal analysis cannot pinpoint exactly when an arrhythmic patient will next experience a rapid heart rate, doctors may be able to use fractal analysis to better understand the patterns within arrhythmia.

Sleep Physiology

Scientists have also used fractal analysis to study the patterns of heartbeat during sleep. Liebovitch and his co-authors note that your heartbeat follows different beat patterns during the three stages of sleep: deep sleep, light sleep and REM sleep. Fractal analysis has shown that only in the period of REM sleep does your brain activity correlate predictably with your heartbeat. Your heartbeat during deep and light sleep is subject to change based on varying brain activity and breathing time, both of which can somewhat be predicted using fractal analysis.

Diagnosis Potential

An emerging field of medicine involves using fractal analysis to diagnose patients who are likely to suffer a heart attack. In "Self-Organized Biological Dynamics and Nonlinear Control," contributing authors Peng, Hausdorff and Goldberger gathered 69 heart rate recordings from both patients with chronic congestive heart failure."

"A healthy heart beats in an aperiodic rhythm, not too regular or repetitive, and not too random or chaotic. The healthy rhythm lives between those extremes, exhibiting a pattern of fractal variability. Loss of fractal variability signals heart disease, pathological dynamics of the heart. In congestive heart failure, for example, the heartbeat is overly regular, corresponding to a low fractal dimension. And in atrial fibrillation, the heartbeat is overly random, corresponding to a high fractal dimension . Many dynamical diseases have this common form, a departure a *loss in a* system's behavior across time."[4]

---

[4] http://www.apa.org/science/about/psa/2007/02/van-ordern.aspx

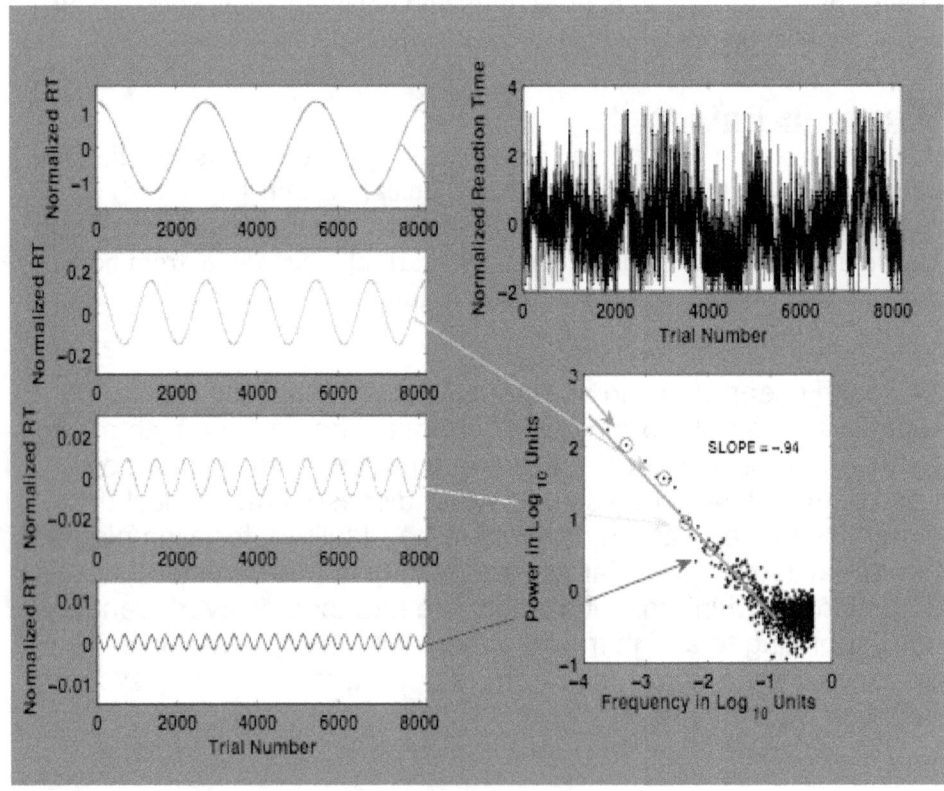

Spectral Portrait of Fractal Behavior

What Do the Fractal Patterns Mean?

"In all likelihood then, fractal behavior tells us about coordination of component processes in the minds and bodies of living organisms. It is tantalizing in this respect that the common fractal signature of healthy functioning is found widely in natural systems that self-organize their behavior, that self-organization actually predicts the ubiquitous fractal signature (e.g., Bak, 1996; Solé & Goodwin, 2001). Self-organization requires a particular kind of interaction to coordinate the processes that must work together. The precise form of this interaction balances competitive and cooperative processes to create an optimally adaptive and flexible working configuration or *critical state*, hence the technical term *self-organized criticality*."[5]

---

[5] http://www.apa.org/science/about/psa/2007/02/van-ordern.aspx

Section C15. Watch Voice Recorder

Modulation disguises the voice and at the same time allows a peozoelectric crystal to operate in a very high pitch encoded sound, thus greatly reducing physical size allowing the device to be used in a wristwatch. This also turns the voice recorder into a message recorder. Very vary small.

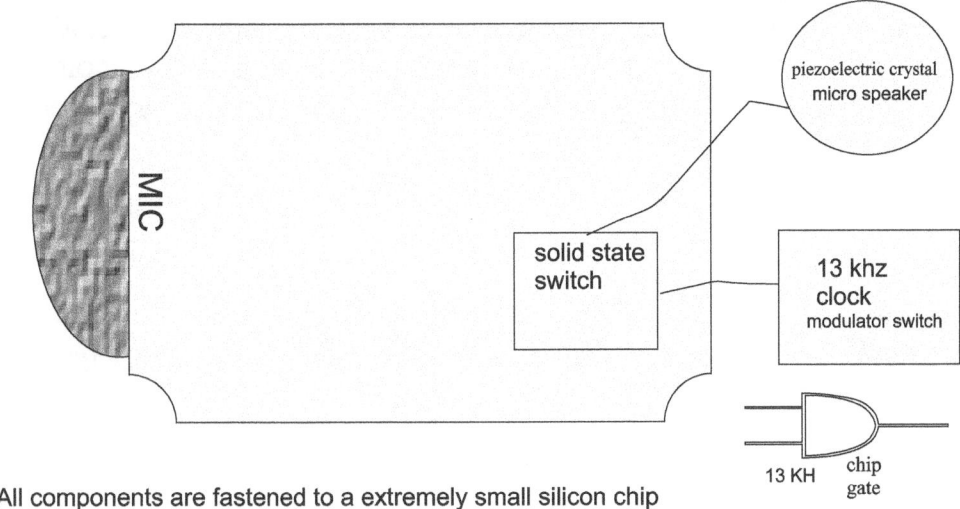

All components are fastened to a extremely small silicon chip

Difference between a message recorder and a voice recorder

A message recorder records the message intended by the user. A voice recorder records the message plus the identity of the user and cannot be used in court against one's permission. Most people do not like to hear the sound of their own voice this makes the message recorder an item that would be in demand because they would not recognize their own voice, and as such would be more likely to purchase this item. People would of course benefit from the small size of the recording of the information. In general people are interested in the message and not in who made the message except with legal matters.

"Piezoelectric speakers are frequently used as beepers in **watches** and other electronic devices, and are sometimes used as tweeters in less-expensive speaker systems, such as computer speakers and portable radios. Piezoelectric speakers have several advantages over conventional loudspeakers: they are resistant to overloads that would normally destroy most high frequency drivers, and they can be used without a crossover due to their electrical properties. There are also disadvantages: some amplifiers can oscillate when driving capacitive loads like most piezoelectrics, which results in distortion or damage to the amplifier. Additionally, their frequency response, in most cases, is inferior to that of other technologies. This is why they are generally used in single frequency (beeper) or non-critical applications."[6]

---

[6] http://en.wikipedia.org/wiki/Loudspeaker

Section C16. Liquid Gyroscope

This describes the concepts of inventions for both a reaction wheel and a gyroscope. For both of these devices we are using the concept of liquid mercury and the Lorenz Force and should never wear out. What we are doing is sending a current from the center of the disc to the edge of the disk inside a pool of liquid Mercury with magnetics on the top or magnetics on the top and bottom. For a reaction wheel we will have one magnet on the top of the disc and the Mercury will sit on the bottom. A current will run from the center of the Mercury to the outside. The material that the Mercury resides in will be insulated and the rim and the center points will be conductive.

For a gyroscope we will essentially have two reaction wheels one on the top and one on the bottom and the Mercury will spin in opposite directions causing the gyroscopic effect. These devices are used in satellite positioning. A reaction wheel will position a satellite however the satellite will continue to move in the direction set by the reaction wheel. In a gyroscope the satellite will be positioned exactly as instructed by the gyroscope. There will be some diagrams and drawings to try and explain how this force works. We will be using the Lorentz force which is basically used as a magnet to propel nonmagnetic conductive materials. We will be using Mercury because it has a density near that of lead and will help us to form a nice reaction wheel and gyroscope.

The spin of a conductive liquid was demonstrated on YouTube when a current was passed from the center of a disc to the outside of a disc.

The difficult part of this concept of invention will be finding a container that can withstand higher pressures under heat, because the Mercury is going to heat up, and the container will expand slightly.

The nice thing about Mercury is that it is a non-viscous liquid and should spin around the desk very rapidly, thus giving a high G-force. There can be no air are gas in the container because when the container is tilted it must exhibit same characteristics as when it's at any angle.

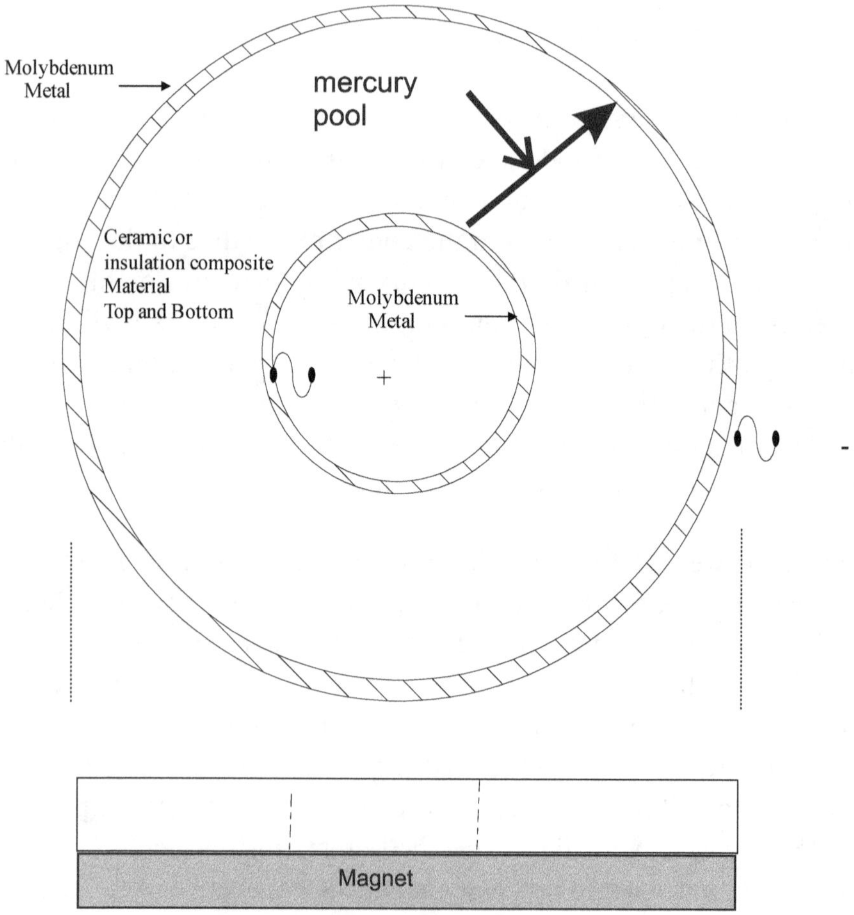

Molybdenum
Metal →

mercury
pool

Ceramic or
insulation composite
Material
Top and Bottom

Molybdenum
Metal →

+

Reaction wheel

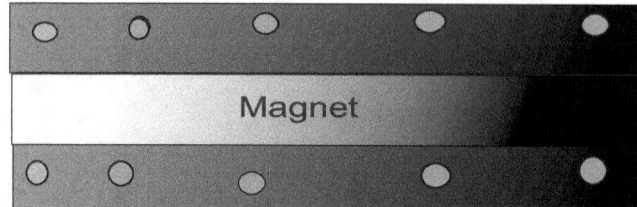

Gyroscope to the left. The magnet is in the middle and the Mercury on top spins one way and the Mercury on the bottom spins another way. There is a hole in the middle of the disk for current conduction.

Magnet

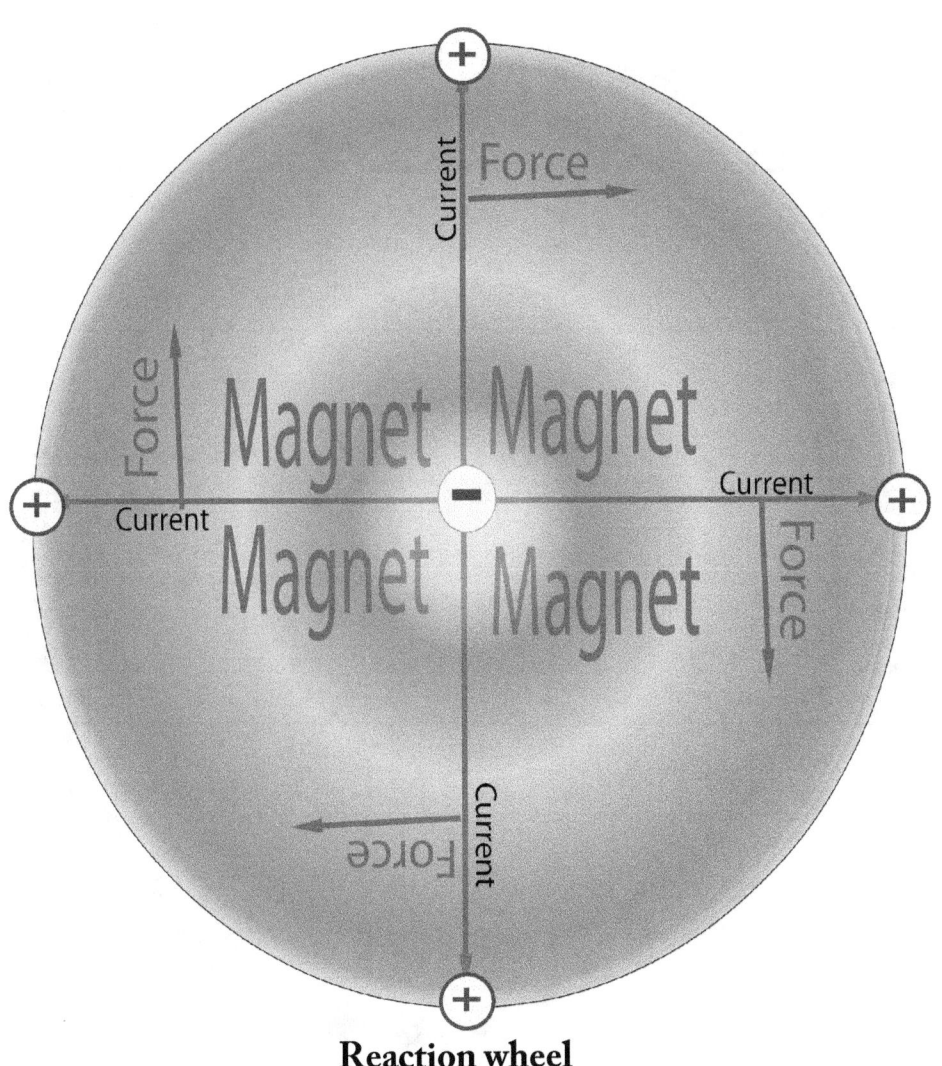

Reaction wheel

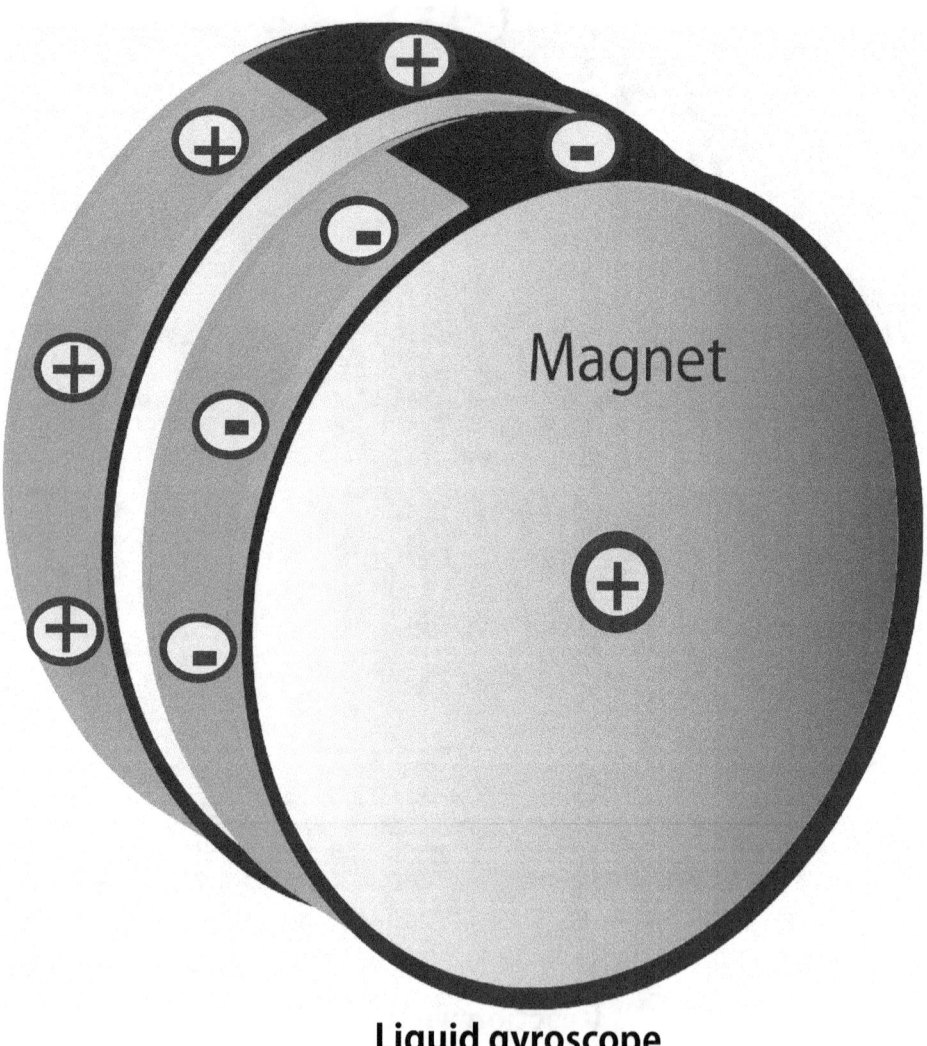

Liquid gyroscope

"*The* magnetic force (q v × B) component of the Lorentz force is responsible for motional electromotive force (or motional EMF), the phenomenon underlying many electrical generators. When a conductor is moved through a magnetic field, the magnetic force tries to push electrons through the wire, and this creates the EMF. The term "motional EMF" is applied to this phenomenon, since the EMF is due to the motion of the wire.

In other electrical generators, the magnets move, while the conductors do not. In this case, the EMF is due to the electric force equation. The electric field in question is created by the changing magnetic field, resulting in an induced EMF, as described by the Maxwell-Faraday equation (one of the four modern Maxwell's equations). "[7]

"Both of these EMF's, despite their different origins, can be described by the same equation, namely, the EMF is the rate of change of magnetic flux through the wire. (This is Faraday's law of induction, see above.) Einstein's theory of special relativity was partially motivated by the desire to better understand this link between the two effects. In fact, the electric and magnetic fields are different faces of the same electromagnetic field, and in moving from one inertial frame to another, the solenoidal vector field portion of the E-field can change in whole or in part to a B-field or vice versa." [8]

"In physics, the Lorentz force is the force on a point charge due to electromagnetic fields. It is given by the following equation in terms of the electric and magnetic fields:

---

[7] http://en.wikipedia.org/wiki/Lorentz_force

[8] http://www.lc.iad.reference.com/browse/electric+force

F = I x B   (force acting in a perpendicular direction)

F is the force (in newtons)

B is the magnetic field (in teslas).

The Lorentz force law has a close relationship with Faraday's law of induction. A positively charged particle will be accelerated in the same linear orientation as the E field, but will curve **perpendicularly** to both the instantaneous velocity vector v and the B field according to the right-hand rule (in detail, if the thumb of the right hand points along v and the index finger along B, then the middle finger points along F)."[9]

"Just 24 days before its third Servicing Mission, the Hubble Space telescope has been placed into a safe mode, triggered by a failure in one of Hubble's last three working gyroscopes.

The safe mode occurred on Saturday 13 November at approximately 14:00 CET (13:00 UT), and this essentially means that the telescope is now in hibernation. This protective safe mode allows ground control of the telescope, but with only two gyros working, Hubble cannot be aimed with the precision necessary for scientific observations of the sky. The Hubble science programme has therefore been suspended until the coming Servicing Mission in early December.

The original six gyroscopes have been the cause of some concern for NASA and ESA officials and astronomers on both continents. The first gyro failed in 1997, the second in 1998, and the third in

---

[9] http://www.numericana.com/answer/maxwell.htm

1999. Even if another gyro should fail in the next few weeks, HST will remain safe. The aperture door has been closed to protect the optics and the spacecraft is aligned to the sun to ensure adequate power is received by its solar panels.

The third Servicing Mission, originally planned for June 2000, was split into two missions - SM3A and SM3B - in part due to the complexity involved, and in part due to the urgency of replacing the failing gyroscopes onboard. It is now clear that it was a very wise decision to speed up the mission, and the fact that HST is now in safe mode stresses even more clearly the importance of SM3A.

The Hubble Space Telescope is a project of international cooperation between NASA and ESA. the next Servicing Mission since two European astronauts will work on fixing Hubble. French astronaut Jean-Francois Clervoy will maneouvre the Space Shuttle's robotic arm during the mission. He will, among other things, have the demanding task of capturing and releasing the Telescope. Claude Nicollier from Switzerland is visiting Hubble for the second time since he flew on the first Servicing Mission. Nicollier will be replacing some of Hubble's parts during some of the mission's spacewalks (also known as EVAs - Extra Vehicular Activities)."[10]

---

[10] http://www.spacetelescope.org/news/heic9904/

"it's difficult to know exactly what went wrong with wheel 4, since the telescope is orbiting far from Earth where no mechanic can reach it. In January, operators noticed elevated friction levels in wheel 4, and those problems continued until April at least. That friction, combined with results from fail tests by the manufacturer, suggests the wheel bearing has failed. "The wheels have this large spinning mass around a nonspinning central hub, and two bearings in between those pieces," Kepler engineer Charles Sobeck told PM. "We think the bearing has fractured. The telescope can operate on just three reaction wheels—a fourth was included as backup. Unfortunately for Kepler, wheel 4 is the second wheel to fail; high levels of friction led NASA operators to shut down reaction **wheel 2 last July.**

Do the wheels have a history of failure?

Reaction wheel 4's demise did not come as a surprise to NASA scientists, because of the previous problems with the wheel. "And we have some history with these wheels by this manufacturer that **they have a limited lifetime,**" former astronaut John Grunsfeld said during last teleconference.

In 2011—two years after Kepler launched—mobile communications company Globalstar ran into a series of **software glitches** in the Goodrich-made reaction wheels on some of its satellites.

Sobeck, however, notes that all machine parts fail at some point, and the wheels, which were manufactured by Goodrich Corp., have had a mixed history of both success and failure. "These same types of wheels are flying on a dozen or more spacecraft," he says, including **Dawn**, ICESat, and TIMED. "Some have operated more than 10 years without any problems. Others fail in orbit, sometimes early, sometimes late."

The spinning components of reaction wheels make them naturally vulnerable. Several NASA satellites have suffered reaction wheel problems, including Dawn, SAMPEX and FUSE, although not all of those spacecraft used Goodrich wheels.

"Two years prior to Kepler's launch, the TIMED satellite had reaction-wheel failure," Sobeck says. After investigating the failure mechanism, the Kepler team implemented a number of design changes that would ensure adequate clearance within the bearing. "We flew a different flavor of wheel than the TIMED flew," he says. "We thought it would mitigate the flaws."

Sobeck says the team considered making more changes before launch, but had neither the time nor resources to continue fiddling with the reaction wheels. "The cost of those changes would have been higher and higher as we went on, because the cost per day was quite high," he says. "At some point it becomes a threat to whether the mission ever gets off the ground." He stands by the team's decision to use the wheels, pointing out that they outlasted the 3.5-year mission they were originally designed for.

Why didn't NASA send extra backups?

Kepler launched with just one backup wheel, even though its orbit took it too far from Earth for a repair mission to be feasible. Why not more? "Every bit of extra redundancy adds extra complexity and extra cost," Sobeck says. Flying four reaction wheels is the industry standard, he added. "I don't know of any spacecraft that uses more or less than four."

What's being done to fix it?

Right now Kepler is being controlled by its thrusters. That uses a lot of fuel and makes the spacecraft unstable. The reaction wheels aimed Kepler so precisely and delicately that the instrument would wiggle by only a millionth of a degree. In contrast, the thrusters create a swing of plus or minus 20 degrees, making Kepler scientists more likely to get seasick than discover new exoplanets.

The team is working to get wheel 2 and wheel 4 running again. By forcing the wheels to spin, the team might be able to remove the fractured elements lodged in the bearings or break them into smaller pieces to free up movement. Even if this option works, the wheels probably will never run as smoothly as they once did. But it would be a major improvement over thruster power.

There's another option that has never been attempted before. Until now, Kepler has had two modes: a thruster-powered safe mode, and a science mode that uses the reaction wheels. Sobeck says the telescope might be able to operate effectively in an in-between state, using its two healthy wheels and only some of the thrusters to keep the telescope properly aimed. The accuracy in this mode wouldn't be as good as before, but it should be considerably better than the current 20-degree margin of error, Sobeck says.

Will Kepler continue finding exoplanets?

That all depends on how accurately Kepler can point after it's been patched up—the degree of error defines the science capability, Sobeck says. It's going to be a couple of months until the engineers know whether Kepler is

salvageable or not."[11]

---

[11]http://www.popularmechanics.com/science/space/telescopes/can-kepler-nasas-planet-hunter-be-saved-15502723

Section C.17 Space Walk Propulsion

This concept of invention device can be used for spacewalking with or without a tether and only requires a small lithium battery. The device uses a compressor type propeller and a non-viscous fluid that creates a high pressure and is directed through tubing that is matched on each side with rough surface area in order that some of the momentum from the fluid is transmitted to the shell of the device and thereby is transmitted to the astronaut or space walker.

Any device that is used now is most likely using a fuel or pressure from a fuel tank. It is believed that with this device has a much longer time that can be spent in space because there will be a sufficient amount of electricity to allow the space walker to remain in space.

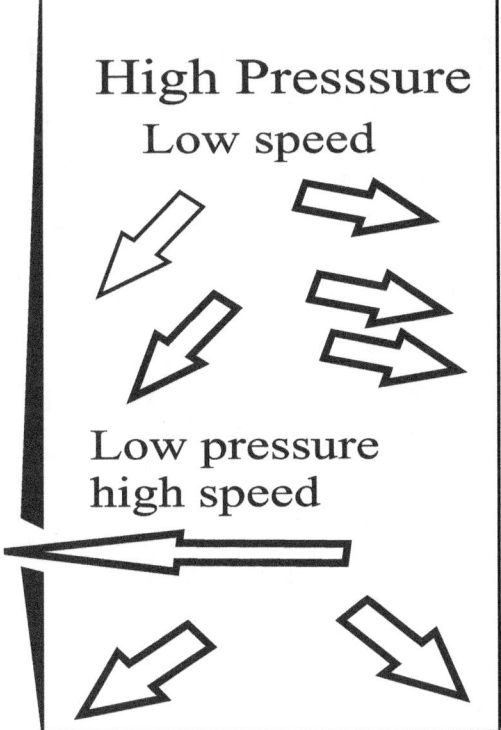

High Presssure
Low speed

Low pressure
high speed

This concept of invention device works by using a non-viscous fluid. By taking the fluid and accelerating at through tubes on each side that are asynchronous some of the momentum of the fluid is transferred to the shell of the device. The figure on the left illustrates that a high pressure system will have no direction whereas a low pressure system will turn the fluid into a propulsive frictional and directional force.

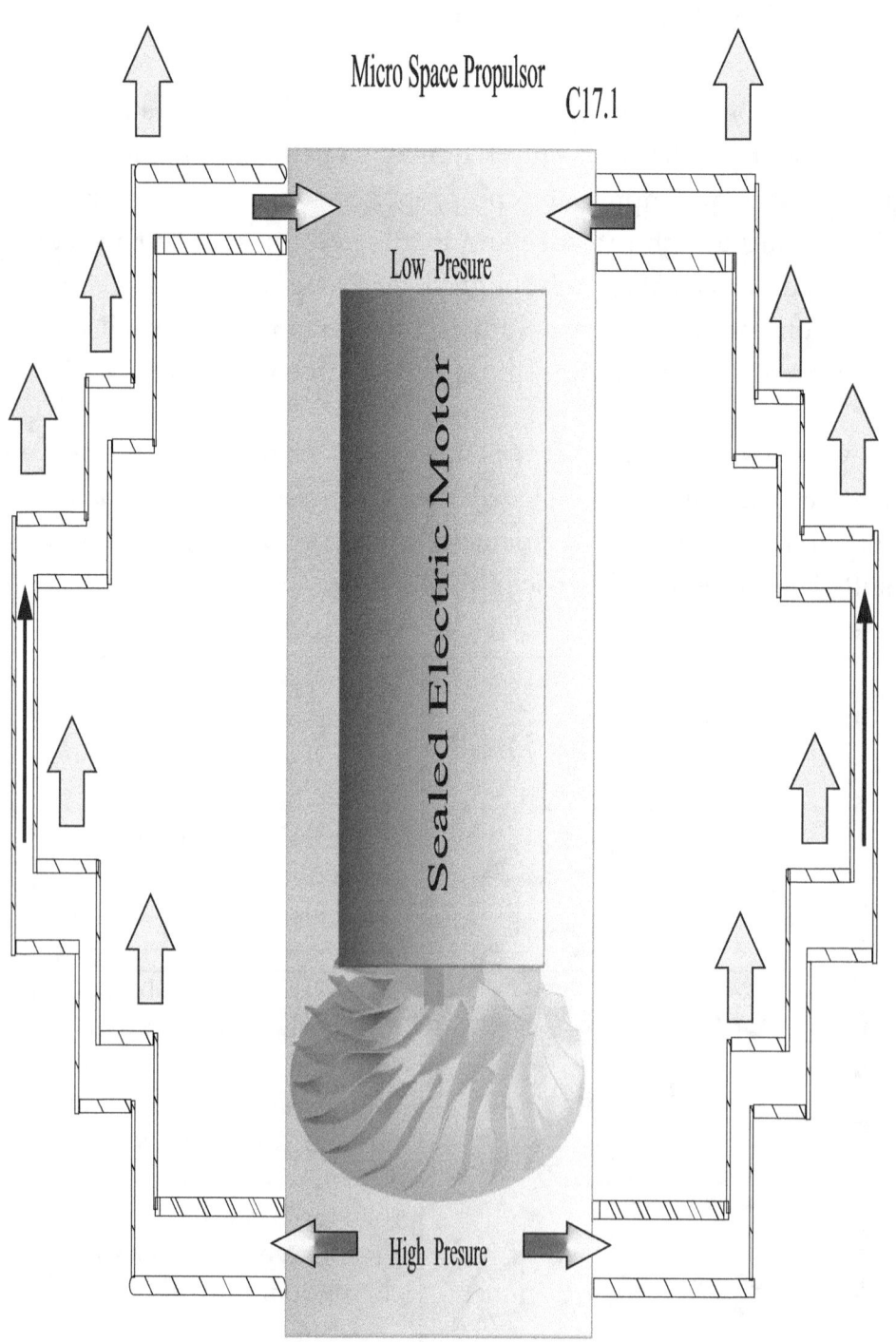

Section C18. **High Powered Pulsing Space Propulsion**

This concept of invention device can be used for relatively high powered space travel. This space travel would only require electricity. The device uses a solenoid pumping action and a non-viscous fluid that creates a high pressure and is directed through tubing that is matched on each side with rough surface area in order that some of the momentum from the fluid is transmitted to the shell of the device and thereby is transmitted to the space capsule. Since the devices are small as many as 20 could be used per capsule and synchronized into a continuous action.

It is believed that with this device a much longer period of time can be spent in space with high power because there will be a sufficient amount of electricity from many various sources in space.

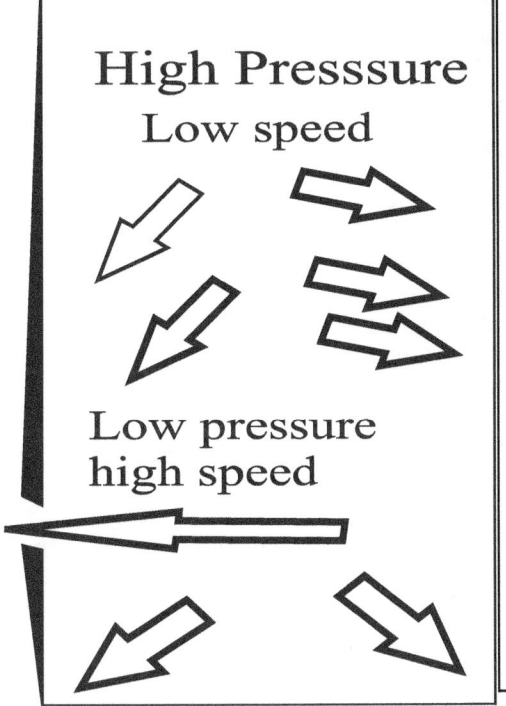

High Presssure
Low speed

Low pressure
high speed

This concept of invention device works by using a non-viscous fluid. By taking the fluid and accelerating at through tubes on each side that are asynchronous some of the momentum of the fluid is transferred to the shell of the device. The figure on the left illustrates that a high pressure system will have no direction whereas a low pressure system will turn the fluid into a propulsive frictional and directional force.

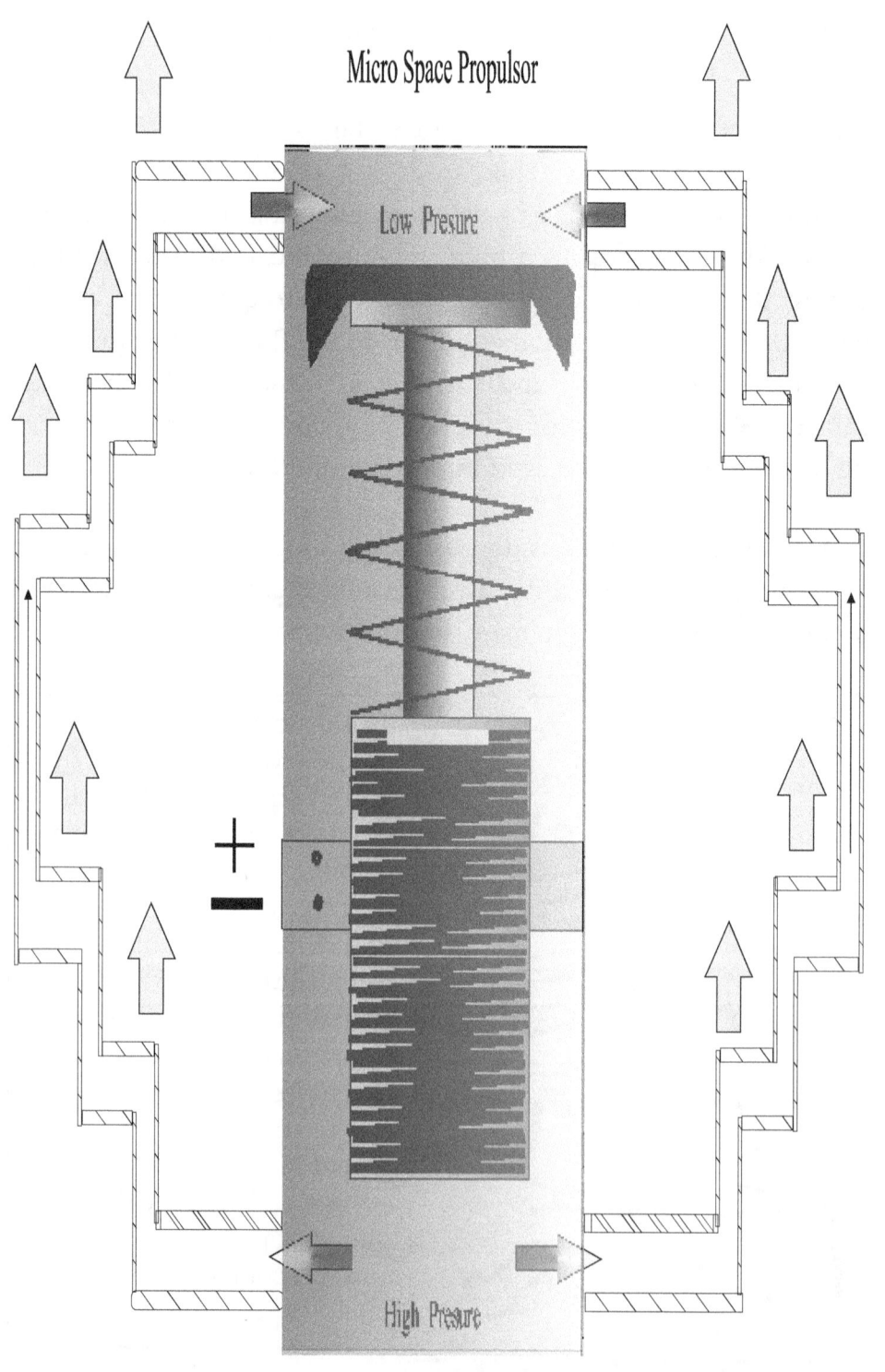

Section C19. Air Conditioned Suit

This concept of invention is designed to be an air conditioned suit for soldiers who fight in very hot climates such as Afghanistan Iran etc. and may also be used for extremely cold climates such as the South Pole. If used as an air conditioned suit the type of material needs have some conductivity of heat. Also the size of the chip must be determined. When used in constructing the suit many chip sizes are available making the construction of the suit more or less logistics. A mylar trace can be run from each chip to chip. For the power source the suit would ideally use a radium salt battery however a regular lithium battery should work. See figure C 19.1.

If used as an air conditioned suit there should be some insulation on the top of the chips to prevent burning. This material may be unique and hard to find as is the material for the suit which must be both rugged and have some heat conduction.

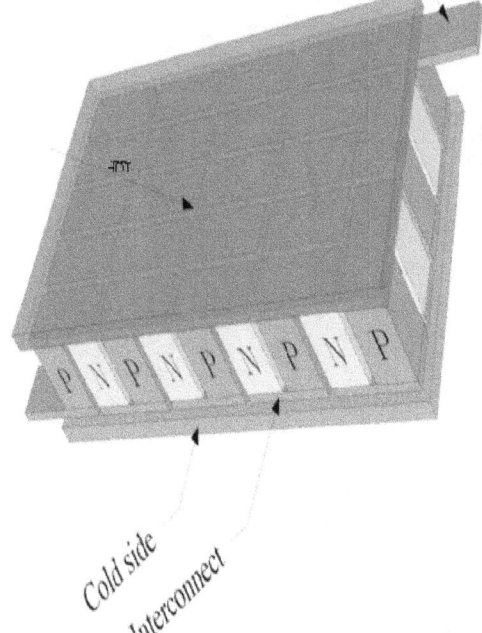

Cold side
Interconnect

A Peltier device or chip is shown on the left. This device can indeed be used to cool people down or heat them up.

The chip has only one input and one output. A current is sent through the chip and one side of the chip gets hot and the other side of the chip gets cold. These chips are actually used for air conditioning systems.

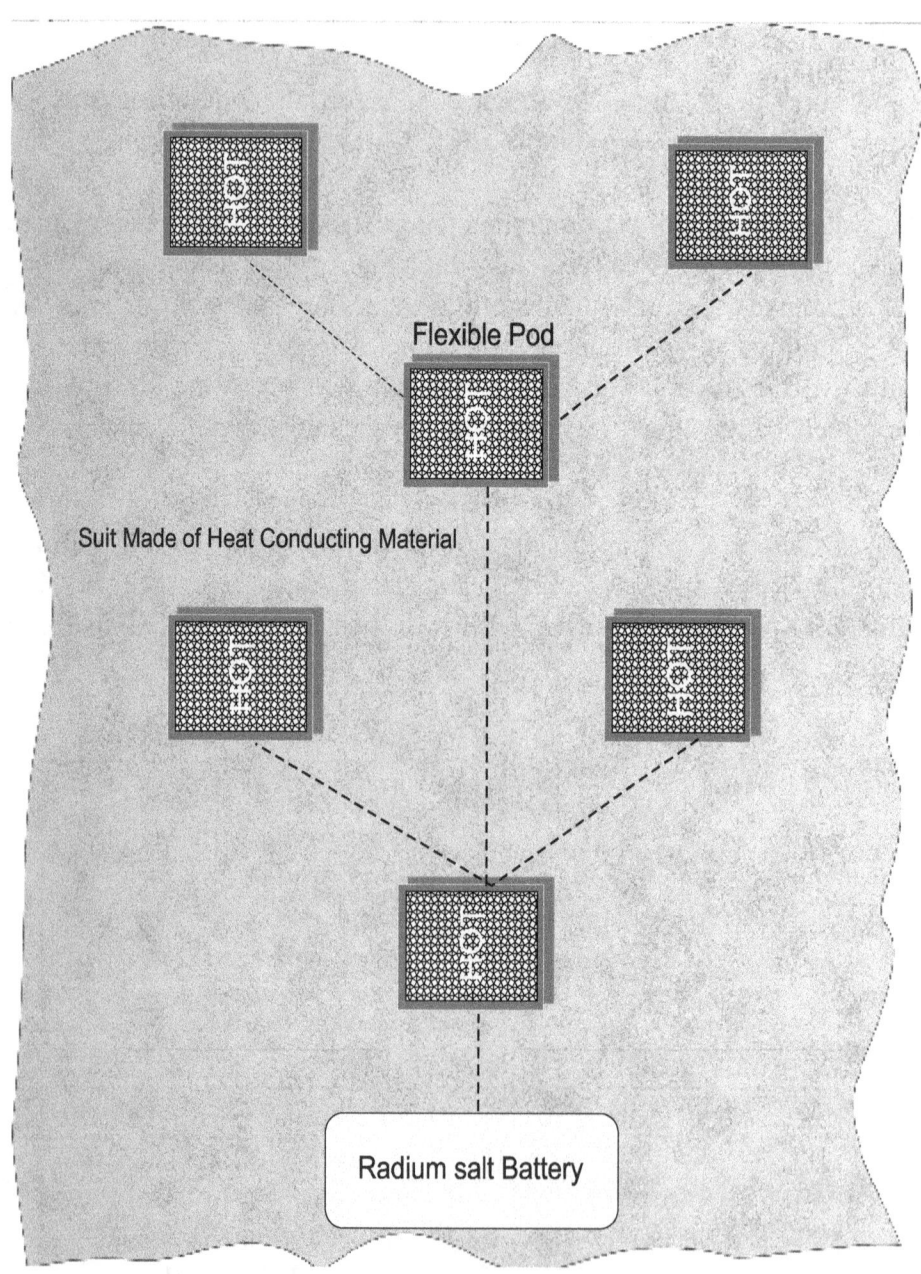

Flexible Pod

Suit Made of Heat Conducting Material

Radium salt Battery

* Chips are hot on one side and cold on the other. These chip are common knowledge and have been available for a while.

The radium battery is not talked about very much , but has been used in the Amy for some time,

C19.1

"Thermoelectric cooling uses the Peltier effect to create a heat flux between the junction of two different types of materials. A Peltier cooler, heater, or thermoelectric heat pump is a solid-state active heat pump which transfers heat from one side of the device to the other, with consumption of electrical energy, depending on the direction of the current. Such an instrument is also called a Peltier device, Peltier heat pump, solid state refrigerator, or thermoelectric cooler (TEC). They can be used either for heating or for cooling (refrigeration), although in practice the main application is cooling. It can also be used as a temperature controller that either heats or cools.

This technology is far less commonly applied to refrigeration than vapor-compression refrigeration is. The main advantages of a Peltier cooler (compared to a vapor-compression refrigerator) are its lack of moving parts or circulating liquid, and its small size and flexible shape (form factor). Its main disadvantage is high cost and poor power efficiency. Many researchers and companies are trying to develop Peltier coolers that are both cheap and efficient. (See Thermoelectric materials.)

A Peltier cooler can also be used as a thermoelectric generator. When operated as a cooler, a voltage is applied across the device, and as a result, a difference in temperature will build up between the two sides. When operated as a generator, one side of the device is heated to a temperature greater than the other side, and as a result, a difference in voltage will build up between the two sides (the Seebeck effect). However, a well-designed Peltier cooler will be a mediocre thermoelectric generator and vice-versa, due to different design and packaging requirements.

Benefits of using **thermoelectric cooling** are:

- No moving parts, fluid, or refrigerants
- **Flexible shape (form factor); in particular, they can have a very small size**
- Has a long life, with mean time between failures (MTBF) exceeding 100,000 hours
- **Is controllable via changing the input voltage/current**

Thermoelectric cooling uses the **Peltier effect** to create a **heat** flux between the junction of two different types of materials. A Peltier cooler, heater, or **thermoelectric** heat pump is a solid-state active **heat pump** which transfers heat from one side of the device to the other, with consumption of **electrical energy**, depending on the direction of the current. Such an instrument is also called a Peltier device, Peltier heat pump, solid state refrigerator, or thermoelectric cooler (TEC). They can be used either for heating or for cooling (refrigeration), **although in practice the main application is cooling**. It can also be used as a temperature controller that either heats or cools.

The main advantages of a Peltier cooler (compared to a vapor-compression refrigerator) are its lack of moving parts or circulating liquid, and its small size and flexible shape (form factor). A Peltier cooler can also be used as a thermoelectric generator. When operated as a cooler, a voltage is applied across the device, and as a result, a difference in temperature will build up between the two sides. When operated as a generator, one side of the device is heated to a temperature greater than the other side, and as a result, a difference in voltage will build up between the two sides (the Seebeck effect)"[12].

"Thermoelectric coolers (TECs), also known as Peltier coolers, are solid-state heat pumps that utilize the Peltier effect to move heat.

Passing a current though a TEC transfers heat from one side to the other, typically producing a heat differential of around 40°C—or as much as 70°C in high-end devices—that can be used to transfer heat from one place to another.

The Peltier Effect

The principle of thermoelectric cooling dates back to the discovery of the Peltier Effect by Jean Peltier in 1834.

All electric current is accompanied by heat current (Joule heating). What Peltier observed was that when electric current passed across the junction of two dissimilar conductors (a "thermocouple") there was a heating effect that could not be explained by Joule

---

[12] http://en.wikipedia.org/wiki/Thermoelectric_generator

heating alone. In fact, depending on the direction of the current, the overall effect could be either heating or cooling. This effect can be harnessed to transfer heat, creating a heater or a cooler.

Peltier himself did not appreciate the potential of his discovery, and it was not efficiently exploited until the end of the 20th century.

How it Works

When two conductors are placed in electric contact, electrons flow out of the one in which the electrons are less bound, into the one where the electrons are more bound.

The reason for this is a difference in the so-called Fermi level between the two conductors. The Fermi level represents the demarcation in energy within the conduction band of a metal, between the energy levels occupied by electrons and those that are unoccupied.

When two conductors with different Fermi levels make contact, electrons flow from the conductor with the higher level, until the change in electrostatic potential brings the two Fermi levels to the same value. (This electrostatic potential is called the contact potential.)

Current passing across the junction results in either a forward or reverse bias, resulting in a temperature gradient.

If the temperature of the hotter junction (heat sink) is kept low by removing the generated heat, the temperature of the cold plate can be cooled by tens of degrees.

Choosing Materials

At first glance metals with their low electrical resistance might seem like a good choice for TEC construction; however they also have high thermal conductivity. This tends to work against any heat gradient produced, and lowers their overall *ZT* value.

In practice semi-conductors are the material of choice. These are usually manufactured by either directional crystallization from a melt or pressed powder metallurgy.

The thermoelectric semiconductor material most often used in today's TE coolers is an alloy of Bismuth Telluride (Bi_2Te_3) that has been suitably doped to provide individual blocks or elements having distinct "N" and "P" characteristics. Other thermoelectric materials include Lead Telluride (PbTe), Silicon Germanium (SiGe), and Bismuth-Antimony (Bi-Sb) alloys, which may be used in specific situations; however, Bismuth Telluride is the best material in most computer cooling scenarios.

Bismuth Telluride has two characteristics worthy of note. Due to its crystal structure, Bismuth Telluride is highly anisotropic. Its electrical resistance is about four times greater parallel to the axis of crystal growth than perpendicular to it. Thermal conductivity, on the other hand, is about double parallel to the crystal-growth axis that perpendicular direction. Hence the anisotropic behavior of resistance is greater than that of thermal conductivity, and the highest Figure-of-Merit occurs in the parallel orientation. The thermoelectric elements must be incorporated into a cooling module so that the crystal growth axis is parallel to the length of each element (perpendicular to the ceramic plates), so that this anisotropy is harnessed for optimum cooling.

Another interesting characteristic of Bismuth Telluride is that Bismuth Telluride (Bi_2Te_3) crystals are made up of hexagonal layers of similar atoms. While alternate layers of Bismuth and Tellurium are held together by strong covalent bonds, adjacent layers of Tellurium are held together only by weak van der Waals bonds. As a result, crystalline Bismuth Telluride cleaves readily along these Tellurium–Tellurium layers (like Mica sheets). Fortunately the cleavage planes generally run parallel to the C-axis, so the material is quite strong when assembled into a thermoelectric cooling module.

TEC Construction

TECs are constructed using two dissimilar semi-conductors, one n-type and the other p-type (they must be different because they need to have different electron densities in order for the effect to work). The two semiconductors are positioned thermally in parallel and joined at one end by a conducting cooling plate (typically of copper or aluminium).

A voltage is applied to the free ends of two different conducting materials, resulting in a flow of electricity through the two semiconductors in series. The flow of DC current across the junction of the two semi-conductors creates a temperature difference. As a result of the temperature difference, Peltier cooling causes heat to be absorbed from the vicinity of the cooling plate, and to move to the other (heat sink) end of the device—see diagram."[13]

[13] http://www.activecool.com/technotes/thermoelectric.html